Technology Portfolio Planning and Management

Practical Concepts and Tools

Technology Portfolio Planning and Management

Practical Concepts and Tools

by

Oliver S. Yu

 Springer

Oliver Yu
Star Strategy Group
Los Altos, CA, USA

Library of Congress Control Number: 2006930112

ISBN-10: 0-387-35446-8 (HB)   ISBN-10: 0-387-35448-4 (e-book)
ISBN-13: 978-0387-35446-0 (HB)   ISBN-13: 978-0387-35448-4 (e-book)

Printed on acid-free paper.

Printed in the United States of America.

9 8 7 6 5 4 3 2 1

springer.com

Preface

In the last decade, technology planning and management has gained increasing attention in the business world and at educational institutions worldwide. In addition to research planning groups in large companies around the globe, many leading organizations have started to establish the position of Chief Technology Officer or equivalent in order to effectively plan and manage the multitude of information and production technologies they constantly acquire and use to facilitate communication, improve productivity, and enhance competitiveness. On the other hand, in a recent survey, more than 40 universities in the United States alone have programs focusing on technology management. At University of California, Berkeley, the Technology Management program claims an enrollment of 1,500 graduate students. However, given all this attention, there is almost no textbook or even reference book available in the market that deals with one of the central issues of technology planning and management: the optimal allocation of resources among technologies to achieve a desired objective.

During my 10-year tenure as the manager for institute-wide research planning at the Electric Power Research Institute (EPRI) and later for 11 years as the director of technology strategies at SRI International, I was able to apply many of the concepts and tools I learned and developed from my academic training in management sciences at Stanford University to assist senior executives in major organizations throughout the world to improve the selection and modification of their technology portfolios to better support the organizational strategies and goals. In 2000, based on my technical knowledge and professional experience, I created a graduate course on technology portfolio planning and management at San Jose State University. The course has been very well received and is now a core course for the MBA-MSE (master of system engineering) dual degree program at San Jose State University. This book represents the formal compilation of my lecture notes. I hope that it will serve not only as a useful textbook for technology management courses, but also as an informative reference for technical professionals actively engaged in the resource allocation decisions for the effective development or application of technologies.

Many people have helped in the preparation of this book. I would like first to thank the College of Business at San Jose State University, especially Associate Dean Lee Jerrell, for the opportunity to create and teach the course on which the book is based. I am grateful to Fred Hillier at Stanford University for introducing me to Springer Publisher. I also want to express my deep appreciation to my former colleagues at EPRI and SRI, including Ali Reza, Charlie Rudasill, Ric Rudman, and Tom Boyce; the many friends I have made during my career, especially Mitsuho Uchida at the Central Research Institute of Electric Power Industry and Hisa Yamamoto at Waseda University; as well as dozens of my students, for their insightful reviews and valuable comments. Finally, I dedicate this book to my dear wife, Joanna, and our children, Christopher and Hamilton.

Oliver S. Yu
Los Altos Hills

Table of Contents

1 Introduction and Overview

Development and application of technologies, which include not only new products and services, but also innovative operation and management techniques, has always been a major force in human history. The invention and use of fire, the wheel, farming tools, gunpowder, the printing press, and workforce organization and specialization are only but a few examples that have revolutionized our civilization. Advances and widespread application of medical, agricultural, military, transportation, and information technologies, as well as breakthroughs in management science concepts and methods, have exerted dominant influences on business organization, economic developments, sociopolitical movements, global ecology, and international powers. As a result, effective planning and management of technology development and application have become critically important, not just to an organization or even a country, but also to the future of mankind as a whole.

1.1. General Scope of Planning and Management of Technology Development and/or Application

As depicted in Figure 1.1, the general scope of planning and management of technology development and/or application can be very broad. It can expand outwardly to include overall corporate business strategies, national economic policies, and even global economic and ecological developments. It can also go inwardly to focus on the planning and management of specific technology projects. Moreover, the planning and management process would involve not only the understanding of how innovative technologies are developed and adopted, but also the assessment of the economic, environmental, sociopolitical, and international impacts of these developments and applications. Such understanding and assessment can then be used to formulate:

- public and government promotional and regulatory policies on technology development and application
- government and business investment strategies on technology development
- business and individual user investment strategies on technology procurement and application
- detailed implementation and management procedures for specific technology projects

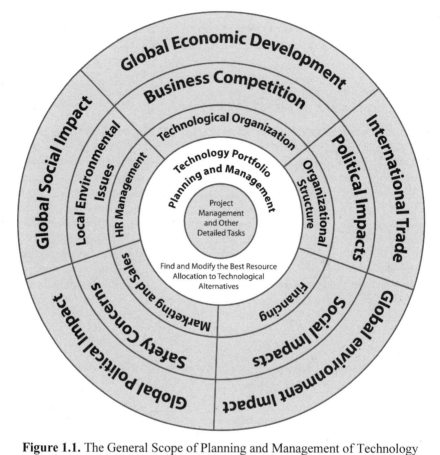

Figure 1.1. The General Scope of Planning and Management of Technology

1.2. Focus, Purpose, and Emphasis of This Book

Of this enormous general scope of the planning and management of technology development and application, this book will *focus on finding and modifying the best technology portfolio, i.e., the best resource allocation among technologies,* to achieve the desired objective of a decision maker. (A portfolio may consist of only a single technology if the best allocation is to devote all resources to one technology rather than spread them among many technologies.) The *purpose* of the book is then *to present practical concepts and tools* for assisting executives as well as analysts in finding (planning) the best portfolio and effectively modifying (managing) it in response to future changes. This portfolio planning and management process can be used by a technology supply as well as a technology application organization. Typical examples are:

- A government agency would like to determine the best allocation of its limited economic, technical, and management resources among existing and prospective technologies to maximize the probability of humans landing on Mars by 2030.

- A corporate research center would like to find the best technology development portfolio within its financial budget and technical capability that would provide a unique competitive edge for the corporation in the next decade.
- A large business office would like to find the best way to invest its limited financial and human resources among the many information and communication technologies to maximize office productivity in the coming year[1].

With the specific focus on resource allocation to technologies, this book will assume that the values, as represented by the goals and objectives of technology portfolio planning and management, are already internally *consistent* with the organization's overall mission and strategy. It will further assume that organizational structure and management capability are in place and ready to implement the best technology portfolio.

Many of the concepts and tools presented in this book, such as the analytic hierarchy process, utility theory, time series forecasting, scenario analysis, linear programming, dynamic programming, decision tree analysis, real options analysis, and the project evaluation and review technique, are well established technical disciplines. The book will emphasize the *basic principles and practical applications,* rather than provide exhaustive coverage, theoretical derivations, or research advances, of the concepts and tools that are particularly relevant to technology portfolio planning and management.

A major challenge in presenting these concepts and tools has been that their applications can be very technology-field specific so that readers of one field, such as information technology, may have little understanding and interest in another field, such as energy technology. To appeal to readers of all fields, the book will emphasize the *use of generic and easy-to-understand application examples.*

1.3. Technology Portfolio Planning and Management Decision Framework

With the specific focus on resource allocation, technology portfolio planning and management is basically the decision process for finding and modifying the best resource allocation among technologies for either development or application purposes by an organization. As depicted in Figure 1.2, the general framework of this decision process has the following key elements:

---

[1] Even though this book concentrates on technology portfolio planning and management for large organizations, the underlying principles of these concepts and tools can apply equally well to individual technology users who nowadays are also making significant investments in the acquisition and application of a variety of household, information, and communication technologies.

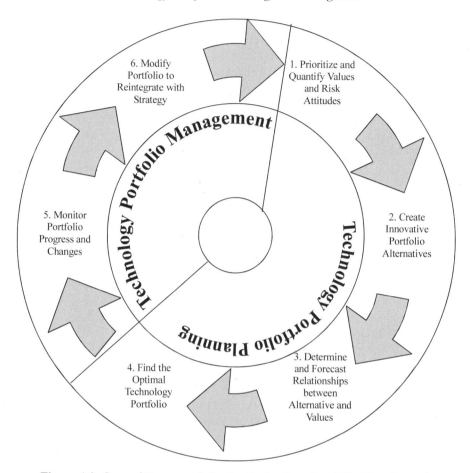

Figure 1.2. General Framework for the Technology Portfolio Planning and Management Decision Process

For technology portfolio planning:

1. **Prioritize and quantify values**: to explicitly prioritize and quantify the values of the decision maker in terms of the degree of relative preference of the various selection criteria for the technology portfolio. The popular positive and negative values include corporate profit or loss, economic growth or decline, and societal benefit and cost. These values are often expressed as the goals and objectives for the portfolio. We assume that the appropriate values for an organization have already been identified based on the overall organizational mission and strategy, but need to be more precisely measured for the technology portfolio planning and management process.

2. **Create innovative portfolio alternatives**: to creatively identify all major relevant, viable, and innovative portfolio alternatives in technology

development or application for the decision maker to choose in order to achieve the optimal values to the organization.

3. **Determine and forecast relationships**: to accurately determine the corresponding relationships between alternatives and values. In the simplest case, these corresponding relationships can be represented by payoff tables for various alternatives, such as the benefit and cost of different resource allocations. In the more sophisticated cases, they are the physical, engineering, socioeconomic models of the impacts of these alternatives on the values. Because we are dealing with future technology impacts on these values, appropriate incorporation of the effects of uncertainty plays an important role in determining these relationships.

4. **Find the optimal technology portfolio**: to apply various innovative methods and tools to find the best resource allocation alternative for technology development or application that will provide the optimal values to the decision maker.

For technology portfolio management:

5. **Monitor portfolio progress and changes**: to closely monitor and track progress of the portfolio and changes in organizational values, available alternatives, and corresponding relationships as technologies and the business environment evolve over time.

6. **Modify the portfolio to reintegrate with organizational strategy**: to reapply various methods and tools to effectively modify the chosen technology portfolio through reallocation of the resources, so that the portfolio can continually maintain an optimal response to changes in values, alternatives, and relationships, and reintegrate itself with any revisions in the organizational strategy.

1.4. Complexity of Technology Portfolio Planning and Management Decisions

These key elements of the technology portfolio decision framework are fraught with complexities and difficulties. For example,

- Values of an organization are often difficult to pinpoint and assess for a number of reasons:
  - The decision maker of an organization is generally an agent (for example, a top government official or business executive) acting on behalf of a principal (such as the citizens or the shareholders). Although the agent should make decisions based on the values of the principal, in reality, there are often conflicts of interest between the agent and principal that could cause the decision maker of the organization to camouflage his or her own true values, which may be the ultimate basis for the decisions.

— Even when the interests of the principal and the agent are aligned, decisions in an organization are usually made by a group, where the participants may still have different perceptions about the importance of their individual functions. Such different perceptions could result in diverse views about the proper values for the organization. For example, technology executives may tend to overemphasize the importance of technology development to the long-term goals and mission of an organization, while production or marketing executives may perceive that their individual functions hold a greater importance for the near-term survival and growth of the organization. How to effectively balance and integrate these diverse perceptions of the real values of the organization poses a major challenge.

— Moreover, values have many intangible aspects, such as prestige and sociopolitical considerations, which are usually difficult to explicitly define and assess, and thus are often inappropriately neglected in the decision process. Furthermore, values are dynamic and shift with time. For example, the values of the decision maker of a start up company may concentrate heavily on survival and growth, while the values of the chief executive of a mature organization may weigh more towards stability and legacy. Thus, the commonly cited profit maximization or cost minimization focus of business decision is often an oversimplification rather than a universal value.

- Identifying innovative alternatives relies heavily on creativity. Without exercising creativity, the decision maker will, often out of fear or ignorance of things unfamiliar, unnecessarily limit him/herself to a set of conventional alternatives while incorrectly excluding many innovative and more effective alternatives. On the other hand, creativity needs to be exercised productively so that resources will not be wasted in evaluating too many impractical alternatives. However, there is no known systematic and foolproof process for developing creativity. Thus, how to creatively identify innovative and effective alternatives without becoming chaotic and counterproductive in the process remains a significant challenge.

- Many disciplines, such as ecology, economics, sociology, and political science, have been devoting enormous effort in attempting to accurately model the relationships between the development and application of a technology and the major short-term and long-term values of an organization, such as business profit, market share growth, organizational prosperity, general economic impact, social equity, and environmental quality. However, unlike the well-established causal relationships in physical laws and engineering disciplines, which have been confirmed through thousands of years of controlled experimentations, there continues to be little validation of the many economic, sociopolitical, and ecological theories and models. As a result, there still is a severe lack of knowledge and understanding of the business, economic, ecological, and political impacts of technology development and application. This lack of knowledge and understanding is made even more serious by the uncertainty created by the dynamic and complex interactions between technology developments and applications and the economy, the society, and the environment at large.

Due to these major complexities of and difficulties in assessing values, identifying alternatives, and determining relationships between alternatives and values, we will start the technology portfolio planning and management decision process with the following simplifying yet reasonable assumptions:

- A single individual or a coherent group who is willing and able to clearly articulate the fundamental values of the organization will act as the sole decision maker for the organization in technology portfolio decisions.
- Major relevant, viable, and innovative alternatives have already been clearly identified.
- Valid relationships between alternatives and values can be readily determined through quantitative analysis or qualitative exploration.

With these assumptions, this book can then concentrate on presenting innovative and effective concepts and tools for quantifying the values, integrating the effects of uncertainty into the relationships, and finding and modifying the optimal portfolio for an organization in its technology portfolio planning and management decisions.

1.5. The Role of Analysis and a Proper Balance

In technology portfolio planning and management, as in other decision processes, analysis based on the concepts and tools in this book plays a key role in quantifying values, assessing alternatives, determining relationships between alternatives and values, and finding and modifying the best choice. Moreover, the general principle of analysis in providing a logical, structured, and the systematic approach to effectively understand the problem and search for the best solution is essential to all decision processes.

Ideally, we would like analysis to help precisely specify the values, comprehensively assess the alternatives, accurately determine the relationships, and pinpoint the best choice. However, the extent of analysis itself is a resource allocation decision problem. The positive values we hope to achieve through analysis in terms of accuracy, precision, and specificity must be properly balanced with the negative values of analysis as reflected by the requirements for financial, human, time, and other resources.

In reality, most decision makers would intuitively undertake a decision process initially through qualitative estimation of the values, quick identification of major alternatives, and simple structuring and categorization of the relationships between alternatives and values, such as payoff tables or priority lists. If the decision is sufficiently complex and important, then a series of more in-depth quantitative analysis may ensue to determine the detailed relationships between alternatives and values to develop a systematic search for the best choice.

Nevertheless, some decision makers tend to err on the side of underanalysis, often due to the dubious perception that the positive values of analysis would not justify its resource requirements. This perception is especially prevalent among decision makers who pride themselves as action-oriented and regard analysis as a sign of weakness and indecisiveness. Unfortunately, a lack of adequate analysis could easily lead to haphazard judgment-based actions without sufficient understanding and reasoning, which produces undesirable results and further confusions when the issues are too complex for intuitive judgments.

On the other hand, we must guard against over-analysis when there is insufficient information, time, or other resources. Specifically, if our understanding of the relationships between alternatives and values is very limited, then development of detailed analytical models of these relationships is not an effective use of the resources. Similarly, if change in technology is rapid, then an elaborate and time-consuming application of analytical tools is again inappropriate. Over-analysis will not only misuse resources, but also can lead to either a false sense of security about incorrect analytical results or a paralysis in decision making by becoming mired in the details.

Even though this book presents many quantitative analysis concepts and tools, we stress the essential *importance of the underlying principles* rather than the detailed mechanics, in providing a logical, structured, and systematic framework for decision making. In reality, many of these principles can be effectively applied conceptually and qualitatively without detailed quantification, unless it is warranted by the value of the decision and the availability of relevant and reliable data. Moreover, this book strives to point out the key assumptions and major advantages, disadvantages, and applicability of these analytical approaches so that a proper balance among the positive and negative values of the analysis can be achieved. In fact, a practical, simple, and balanced approach for analysis as well as for other resource allocation decisions is the well-known Pareto's Law[2], which empirically estimates that about 80% of the resources should be allocated to the analysis of the top 20% of the alternatives or issues.

Another important issue related to analysis is the *subjective nature of decision making.* This inherently subjective nature of decision making sometimes can be perplexing to decision makers as well as analysts with strong quantitative background and training, as they tend to psychologically and semantically associate subjectivity with being closeminded, arbitrary, and random, while preferring to view the ideal decision making process as objective and precise. However, this subjective nature by no means implies any disregard of objective information and analysis. It simply indicates the fact that given all the objective information and analysis, the final choice taken in a decision is ultimately the result of the decision maker's *subjective* perception of the values, awareness of the alternatives, understanding of the relationships between alternatives and values, and ability to search for the best alternative. Any external inputs, such as different perceptions of values, added awareness of alternatives, varying understanding of relationships, and new ability for finding the best alternative, can only have impact on the decision when they are *subjectively* accepted and adopted by the decision maker. On the other hand, an effective decision maker should certainly be earnest and open-minded in collecting and examining a wide range of inputs and be as rigorous as practical in applying analytical concepts and tools for assessing values, creating alternatives, understanding relationships, and finding the best alternative.

---

[2] The author thanks Richard Rudman for pointing out the applications of Pareto's Law to analysis.

1.6. Organization of the Book

Based on the specific focus and simplifying assumptions discussed above, the remainder of this book is organized into the following chapters:

Chapter 2. Quantify Values and Risk Attitudes

This chapter presents several major concepts and tools for quantitatively assessing the values and risk attitudes of the decision maker for technology portfolio planning and management. They include the simultaneous rating method, the analytic hierarchy process, and the utility theory. The values to be assessed and quantified by these tools will include both the monetary value and other less tangible values, such as environmental quality and social-political considerations.

Chapter 3. Create Innovative Alternatives

Although we have assumed that the relevant, viable, and innovative alternatives to be considered for the technology portfolio have already been identified, it is still useful to discuss some of the creative approaches for identifying these alternatives, as such approaches are also important and useful for determining relationships and finding the best portfolio. These approaches include productive brainstorming, lateral thinking, and the theory of inventive problem solving.

Chapter 4. Determine and Forecast Relationships—Major Analytic Methods

This chapter presents the general characteristics, the underlying assumptions, as well as the major advantages, disadvantages, and applicability of the most widely used analytic methods for determining and forecasting the future relationships between technology alternatives and organizational values. These methods include, among others, time series analysis as well as saturation, substitution, and seasonal fluctuation models.

Chapter 5. Determine and Forecast Relationships—Qualitative Approaches

Because of the general lack of accurate understanding of future relationships between technology alternatives and organization values, which is the foundation of the analytic methods presented in Chapter 4, this chapter presents a number of qualitative approaches for forecasting these relationships to supplement the analytic methods. These approaches include the Delphi method, the cross-impact matrix, and particularly, the decision-focused scenario analysis. We discuss the general characteristics, underlying assumptions, as well as major advantages, disadvantages, and applicability of each of these approaches.

Chapter 6. Search for Optimal Choice—Deterministic Approaches

In this chapter, we present basic approaches for finding the best technology portfolio in a deterministic environment, where the impact of technology alternatives on the decision maker's values is either known for certain or the effects of uncertainty are incorporated deterministically into the relationships through the use of expected values. The applicability of these approaches is heavily dependent on the characteristics of the functional form of the value-alternative relationship as well as the set of feasible alternatives. We have thus divided the approaches according to the functional form of the relationships into three groups:

1. those associated with relationships represented by aggregated values, such as a simple payoff table; the approaches include the benefit-cost ratio analysis, the multi-factor evaluation method, and the analytic hierarchy process;
2. those associated with relationships represented by a concave or linear value function and a convex set of alternatives, such as a proportional or diminishing return function from increasing resource allocation; the approaches include the generalized LaGrange multiplier method and linear programming;
3. those provide efficient procedures for finding the best alternative among a large number of subdivisible alternatives; they include the branch-and-bound approach and the dynamic programming approach.

Chapter 7. Search for the Optimal Choice—Decision Under Uncertainty

This chapter presents major concepts and tools for finding the optimal technology portfolio with explicit consideration of the effects of uncertainty in the business environment. These concepts and tools include decision tree analysis, portfolio diversification, and real options analysis.

Chapter 8. Monitor Portfolio Progress and Environmental Changes

In this chapter, we will present two major tools for tracking the progress of projects in the technology portfolio: a classical method for technology project management, the project evaluation and review technique (PERT), and an artificial intelligence-based computer-assisted project portfolio monitoring system, which can also be used to identifying early warning signs of changes in the business environment.

Chapter 9. Modify Portfolio to Reintegrate with Organizational Strategy

In this final chapter, we will present two major qualitative tools for effectively modifying the technology portfolio so that it will continue to support the organizational strategy in a changing business environment. These tools are: Factor Analysis for analyzing the strengths and weaknesses of a technology portfolio; and Strategy Map for checking the portfolio's consistency with different business environment scenarios and changing organizational strategy.

Appendix

In the appendix, we summarize the basic mathematical, probabilistic, and statistical concepts and tools that will be needed to support the decision making process for technology portfolio planning and management.

1.7. References

[1]   Burgelman, R., C. Christensen, and S. Wheelwright, *Strategic Management of Technology and Innovation*, 4th ed., McGraw-Hill/Irwin, 2003

[2]   Cardullo, M., *Introduction to Managing Technology*, Wiley, 1996

[3]   Frankel, E., *Management of Technological Change*, Springer, 1990

[4]   Khalil, T., R. Mason, and L. Lefebvre, *Management of Technology*, Pergamon, 2001

[5]   Linstone, H., and I. Mitroff, *The Challenge of the 21st Century: Managing Our Technologies and Ourselves in a Shrinking World*, State University of New York Press, 1994

[6]   Yu, O., G. Hsu, and T. Chen, *Introduction to Technology Management* (in Chinese), Wu-Nan Publications, 1998, 3rd printing 2003

1.8. Exercises

Problem 1.1

Identify a non-trivial technology portfolio decision that you know or may be involved in your business or personal life (for example, finding the best combination of computer, communication, and entertainment technologies for your home). Describe the key elements of your decision process according to those in the decision framework presented in Section 1.3 of this chapter. (Note that it is possible for a portfolio to contain only one technology.)

Problem 1.2

For the technology portfolio decision in Problem 1.1, describe some of the major difficulties you may have encountered in specifying the values, identifying the alternatives, determining the relationships between alternatives and values, and finding the best portfolio.

Problem 1.3

Describe some of the major economic and sociopolitical impacts of well-known technology developments and applications in the past (for examples, the developments and applications of automobiles, insecticides, nuclear weapons, personal computers, and biotechnologies).

Problem 1.4

In purchasing a new computer, assume that your major values are Affordability, Quality, and Performance. Develop a systematic procedure for using these values to select the best computer for you among three top choices. Demonstrate the procedure with a specific example. The main purpose of this problem is to have a description of your own initial procedure for finding the best choice so that you can compare it with the many procedures presented later in the book.

2 Quantify Values and Risk Attitudes

As in any decision process, the starting point of technology portfolio planning and management is the assessment of the *subjective* values of the decision maker. These values are also often expressed in terms of goals and objectives for the technology portfolio decision as they represent what the decision maker would like to achieve with the portfolio.

As discussed in Chapter 1, eliciting and articulating values are usually difficult. In many cases, because values are complex, multidimensional, emotion-laden, and often evolving, a decision maker may not even be fully aware of his or her true values. In other cases, due to the potential conflict of interests between the decision maker and other stakeholders in the decision, the decision maker may be reluctant to reveal his or her own values. Although highly interesting, the value extraction process is beyond the scope of this book. We shall assume that the decision maker is both willing and able to clearly identify and define his or her subjective values for the technology portfolio planning and management decision.

A technology generally possesses many different values, represented by the characteristics or attributes of the technology that are desirable to the decision maker. Furthermore, the values of individual component technologies in a portfolio need to be integrated into an overall value for the portfolio, so that these overall values of different portfolios can be systematically compared. Thus, once the desired values of the decision maker have been identified and defined, they will still need to be assessed, quantitatively if possible, to make the process of choosing the best alternative, i.e., the resource allocation among these technologies that will yield the optimal overall value, more explicit and definitive. This quantitative assessment of values is especially important in integrating many less tangible values, such as prestige, good will, and other social considerations, together with the more directly measurable financial return and physical output values, such as cost, profit, and production quantity, into an overall value for a technology portfolio.

In this chapter, we will discuss three widely applied approaches for the quantitative assessment of the decision maker's values. All these approaches are based on the common observation and assumption that the levels of importance of different values, as often represented by the criteria or outcomes of alternative technology choices, are reflected in and can be quantified through the degrees of *subjective preference* of these values by the decision maker.

2.1. Simultaneous Rating Approach

The simplest approach is to simultaneously compare and rate the degrees of subjective preference of various values by the decision maker. This can be done through either setting an ordinal rating of the values or by further converting the ordinal ratings into semiquantitative measures. As a specific example, assume that the three major values for a prospective technology portfolio are profitability, quality, and prestige. These values can be further defined as follows:

Profitability: The expected present value of the net profit achieved from the development or application of the technology portfolio in the next 10 years

Quality: The precision, reliability, and durability of the technologies in the portfolio

Prestige: Recognition of the innovativeness of the technologies and impact on the general reputation of the company in the business community

Among these values, only the profitability may be readily quantified, for example, by assuming that the degree of preference is proportional to the amount of expected profitability. Even there, the quantification process may be complicated, as the financial return of the technology portfolio could involve many assumptions and relationships, such as the accounting procedure used and the sequence of cash flows over time. Furthermore, in reality, the degree of preference by the decision maker may not be directly proportional to the amount of expected profitability, which adds to the complexity of estimation. Thus, a qualitative rating is often used to measure the degrees of preference of these values.

The typical qualitative rating measures are Low (L), Medium (M), and High (H). They can be further elaborated into many more ordinal ratings, such as LL, LM, LH, ML, MM, MH, HL, HM, and HH. These ordinal ratings can then be converted into numerical measures, such as 1 for LL, 2 for LM, ..., and 9 for HH. There are also many other semi-quantitative rating scales. The popular ones are: 1 through 3, 1 through 5, 1 through 10, and 1 through 100.

In the above example, the decision maker may rate qualitatively the degrees of preference for profitability as High, for quality as Medium, and for prestige as Low, and assign numerical measures 9, 6, and 3 to these degrees of preference, respectively.

The advantages of this approach are in its simplicity and intuitive appeal. The main disadvantage is in its imprecision, which rises largely from the inherent inability for humans to precisely differentiate several values at the same time.

2.2. Analytic Hierarchy Process

The analytic hierarchy process (AHP) was developed by Thomas Saaty in the early 1970s based on the following observations of the relative preference of values by a decision maker:

(a) Various types of values in a decision process can be divided into different levels or hierarchies of details, so that the degrees of relative preference of the values within a hierarchy by the decision maker are all within the same order of magnitude, i.e., within a factor of 10 of one another.

(b) Human judgments can best differentiate the degrees of relative preference of the values within a hierarchy through pair-wise comparisons of these values

(c) The overall degrees of preferences of these values and their consistency can be assessed from the pair-wise comparisons of these values through the use of matrix theory.

Using AHP to assess values takes the following steps:

(1) Set up the Hierarchy of Values

In this first step, we will classify the values into various levels of detail, so that all values that are comparable within the same order of magnitude, i.e., within a factor of 10, in their degrees of relative preference by the decision maker, belong to the same level of detail or hierarchy.

As a specific example shown in Figure 2.1, the Profitability, Quality, and Prestige values of a prospective technology portfolio can generally be regarded to be in the same order of magnitude, and thus the same hierarchy. On the other hand, Revenues and Expenses as components of Profitability value are not in the same order of magnitude in terms of the level of detail as those of Quality and Prestige values. In this case, the higher hierarchy values can be decomposed into lower hierarchy values that are comparable within the same order of magnitude. For example, Quality value of a technology may be decomposed into lower hierarchy values of Precision, Reliability, and Durability, which are then within the same order of magnitude in the degree of relative preference as that of Revenues and Expenses.

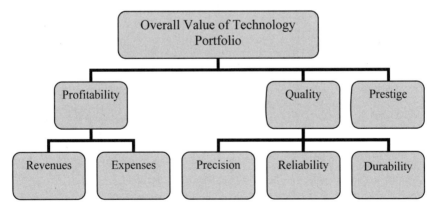

Figure 2.1. Example of the Hierarchy of Values for a Technology Portfolio

The number of hierarchies of values is judgmentally determined to provide sufficient detail so that meaningful and effective pair-wise comparisons can be made for values within a hierarchy.

(2) Set up a Standard Scale for the Pair-wise Comparison

Saaty has set up the following standard scale for the pair-wise comparison, w_{ij}, of the degrees of preference of Value i relative to that of Value j by the decision maker. Specifically, w_{ij} will be:

> 1 — if Value i is equally preferred to Value j
> 3 — if Value i is moderately preferred to Value j
> 5 — if Value i is strongly preferred to Value j
> 7 — if Value i is very strongly preferred to Value j
> 9 — if Value i is extremely strongly preferred to Value j

Even-numbered comparison measures 2, 4, 6, and 8 can be used for w_{ij} lying in between these odd-numbered measures. For example, w_{ij} would be 4 if Value i is moderately to strongly preferred to Value j.

(3) Develop an nxn Pair-wise Comparison Matrix W for the n Values in a Hierarchy.

In this step, the decision maker will use judgments to estimate w_{ij}, the relative preference of Value i to Value j in a given hierarchy. Note that $w_{ii} = 1$ for all i, as it is Value i compared to itself. Furthermore, by definition, $w_{ij} = 1/w_{ji}$. In other words, the relative preference of Value i to Value j is the *reciprocal* of the relative preference of Value j to Value i. Thus, out of the nxn elements in the pair-wise comparison matrix, the decision maker needs only to estimate the relative preferences of $n(n-1)/2$ pairs of values that are neither the diagonal elements nor the reciprocals of each other.

    As an example, for the three values of a prospective technology portfolio, the decision maker needs only estimate by judgment the relative preference through 3 pair-wise comparisons of these values that are not reciprocals of one another. In this specific example, the decision maker may judge that:

- Profitability (Value 1) is moderately to strongly preferred to Quality (Value 2); $w_{12} = 4$
- Profitability (Value 1) is extremely preferred to Prestige (Value 3); $w_{13} = 9$
- Quality (Value 2) is moderately to strongly preferred to Prestige (Value 3); $w_{23} = 4$

Then the 3x3 pair-wise comparison matrix, $W=[w_{ij}]$, would be:

	Profitability	Quality	Prestige
Profitability	1	4	9
Quality	1/4	1	4
Prestige	1/9	1/4	1

(4) Estimate the Average Preferences or Weights of the n Values in a Hierarchy

Notice that w_{ij}'s in the jth column are the relative preferences of Values, $i = 1,2,..,n$ to Value j. Thus, the ratios, $w_{ij}/\Sigma_k w_{kj}$, represents the normalized relative preference of Value i by the decision maker using the relative preference of Value j as the base. As an example, for $i = 2$ and $j = 1,2,3$, $w_{ij}/\Sigma_k w_{kj}$ are $\frac{1}{4}$ /(1+ $\frac{1}{4}$ +1/9)=0.1837, 1/(4+1+ $\frac{1}{4}$)=0.1905, and 4/(9+4+1)=0.2857 respectively. If the decision maker is totally consistent in these pair-wise comparisons, in that $w_{kj} = w_{ki}$ (w_{ij}) for any i, j, and k, then the normalized relative preferences of Value i, $w_{ij}/\Sigma_k w_{kj}$, will all be identically equal to the constant ratio, $1/\Sigma_k w_{ki}$, for all j. However, because of fluctuations in judgments particular when the relative preferences for the values are close to one another, these pair-wise comparisons are often not totally consistent, such as the case in the example. Since we do not know which comparison causes the inconsistency, these ratios for different j's are simply different estimates of the true normalized relative preference for Value i. To smooth out the errors introduced by inconsistency, we will use the row average of the ratios by summing the ratios for each row i and divide them by n as the best estimate for the true normalized relative preference for Value i. The computational formula is given below.

Normalized Relative Preference of Value i = RAV_i = $\Sigma_j(w_{ij}/\Sigma_k w_{kj})/n$ for $i = 1, 2,\ldots, n$.

For the above example, we have

	Profitability	Quality	Prestige	RAV=(row sum)/n
Profitability	1/(1+1/4+1/9)	4/(4+1+¼)	9/(9+4+1)	0.71
Quality	1/4/(1+1/4+1/9)	1/(4+1+¼)	4/(9+4+1)	0.22
Prestige	(1/9)/(1+1/4+1/9)	(1/4)/(4+1+¼)	1/(9+4+1)	0.07

(5) Check Matrix Consistency

As discussed earlier, human judgments can often be inconsistent, especially when the decision maker feels ambivalent about two values. As a result, the decision maker may, for example, prefer Value i to Value j and Value j to Value k in one set of comparisons and then contradictorily prefers Value k to Value i in another set of comparisons. It is thus important to check the consistency of *each* pair-wise comparison matrix.

Let W be the $n \times n$ comparison matrix, $RAV = (RAV_i)$ be the $n \times 1$ column vector of row averages, and $RRAV^T$ be the $1 \times n$ row vector (with the superscript T signifying the transpose of the column vector $RRAV$, which changes it into a row vector), where the ith component is the reciprocals of RAV_i, the ith component of RAV. Then based on matrix theory, the consistency of W can be checked by first estimating the largest eigenvalue λ_{max} of the matrix through the formula below:

$$\lambda_{max} = RRAV^T (W) (RAV)/n$$

If W is *totally* consistent, then $w_{ik}=w_{ij}(w_{jk})$ for all i, j, and k. As a result columns (as well as rows) must be *proportional to one another*, i.e., any columns i and j, $w_{ik}= c(w_{jk})$ for k=1,2,...,n with c= w_{ij} being the proportiona constant. In this case, it can be easily shown that λ_{max} = n. The consistency of W ca then be assessed by the consistency index CI, which measures the deviation of the estimated λ_{max} from n as follows:

$$CI = |(\lambda_{max}-n)/(n-1)|$$

It has been determined *empirically* that if CI ≤ 0.05[3], then W is sufficiently consistent in that there is no significant contradiction among the relative preferences of the values and the ratios $w_{ij}/\Sigma_k w_{kj}$ for each column j are close to each other as well as to RAV_i, the average normalized relative preference of Value i. Otherwise, W is sufficiently inconsistent and revisions will need to be made for the pair-wise comparisons. For the above example, we have

$$RAV^T = (0.71, 0.22, 0.07)$$

$$W(RAV)=[0.71(1)+0.22(4)+0.07(9), 0.71(1/4)+0.22(1)+0.07(4),$$
$$0.71(1/9)+0.22(1/4)+0.07(1)]^T$$

$$RRAV^T= (1/0.71, 1/0.22, 1/0.07)$$

$$CI = \{[(1/0.71, 1/0.22, 1/0.07)W(RAV)/3]-3\}/(3-1) = 0.021 < 0.05$$

Thus, W is sufficiently consistent.

(6) Revise the pair-wise comparison matrix for consistency[4]

If the pair-wise comparison matrix W turns out to be insufficiently consistent, then it can be revised to total consistency through the following procedure:

(a) Let C_0 be the set of n diagonal elements, w_{ii}'s of W.

(b) Let w_{ij} and $1/w_{ij}$, with j>i, be the first off-diagonal reciprocal pair of elements in W of which the decision maker is *most* confident. Then $C_1= \{C_0, w_{ij}, 1/w_{ij}\}$.

(c) In the ith iteration, let w_{hk} and $1/w_{hk}$ be the next most confident reciprocal pair not yet in C_i. If C_i already contains w_{fk}, w_{hg}, and w_{fg}, (note that one of these may be a diagonal element) for some f and g, so that a rectangle or square formed by w_{hk}, w_{hk}, w_{hk}, and w_{hk} exists in the matrix, then to be consistent, w_{hk} must either equal to or be revised to $w_{hg}(w_{fk}/w_{fg})$. Otherwise, w_{hk} remains unchanged and $C_{i+1}=\{C_i, w_{hk}, 1/w_{hk}\}$.

---

[3] Saaty has used empirically determined requirements on CI that vary with n; however, CI$\leq$0.05 is a stronger requirement that satisfies Saaty's other requirements for all values of n.

[4] This section is based on original work by the author.

(d) Continue until all reciprocal pairs have been included and thus all columns become proportional to one another.

Specifically, in the above example, suppose that the decision maker desires to make the pair-wise comparison matrix totally consistent, and he or she is most confident about $w_{12}=4$, second most confident about $w_{13}=9$, and least confident about $w_{23}=4$. Then following the above procedure, we have:

$$C_0 = \{ w_{11}=1, w_{22}=1, w_{33}=1 \}$$

$$C_1 = \{ w_{11}=1, w_{22}=1, w_{33}=1, w_{12}=4, w_{21}= \tfrac{1}{4} \}$$

Since w_{13} does not form a rectangle or square with any elements in C_1, it does not need to be revised and

$$C_2 = \{ w_{11}=1, w_{22}=1, w_{33}=1, w_{12}=4, w_{21}= \tfrac{1}{4}, w_{13}=9, w_{31}=1/9 \}$$

Finally, since w_{23} forms a square in the matrix with w_{22}, w_{13}, and w_{12} in C_2 (in this case, h=2, k=3, f=1, and g=2.), it needs to be revised to $w_{22}(w_{13}/w_{12}) = 9/4 = 2.25$. With this change, the matrix takes the following revised form:

	Profitability	Quality	Prestige
Profitability	1	4	9
Quality	1/4	1	9/4
Prestige	1/9	4/9	1

As a result, all columns of the matrix become proportional to one another. Then CI=0, and the matrix becomes totally consistent.

(7) Distribute the Relative Preference of a Value to Values in a Sub-hierarchy

If value V_i in a given hierarchy can be decomposed into a set of component values, V_{i1}, V_{i2}, ..., V_{iN}, in a sub-hierarchy, then the relative preference of V_i should be distributed to all component values by multiplying it to the relative preferences of the component values obtained from the pair-wise comparison matrix of V_{ik}'s.

In the above example, Quality has a relative preference of 0.22 and is decomposed into three component values: Precision, Reliability, and Durability. Assume that by pair-wise comparison matrix analysis we have obtained the relative preferences (i.e., the RAVs) of these component values as 0.2, 0.5, and 0.3 respectively. Then the overall relative preference for these component values would be 0.22x0.2=0.044, 0.22x0.5=0.11, and 0.22x0.3=0.066 respectively.

Major Advantages and Disadvantages

The major advantages of AHP are:

1. It provides a psychologically sound basis for making a more precise assessment of values through hierarchical structuring and pair-wise comparisons of values within the same hierarchy.
2. It provides a mathematically sound basis for checking the consistency of human judgments.
3. It is simple, intuitive, and easily programmable on a computer
4. The weights or relative importance of the values resulting from the analysis are numerically stable for small inconsistencies in human judgments in pair-wise comparisons
5. In addition to value assessment, it can also be used for forecasting, as well as alternative selection and resource allocation as to be discussed in Chapters 3 and 4, respectively
6. The method has been widely applied and accepted by major business corporations and government agencies throughout the world

On the other hand, AHP has the following pitfalls:

1. The pair-wise comparison may be distorted by human perception. For example, human judgments of light intensity at different distances have been shown to be inconsistent with the physical law that the intensity diminishes with the square of the distance. This may also be caused by the fact that the 1–9 comparison scale is not numerically proportional. For example, although in the preference scale, 5 lies at the exact mid-point of 1 and 9, but numerically the ratio of 5 to 1 is much more significant than the ratio of 9 to 5, which could cause distortions in the final comparison results.
2. If the values to be compared are highly correlated, the comparison results can also be distorted. As an extreme example, if two of the three values are totally correlated, then they should be combined into a single value. By being two separate values, the same value would be over-rated in the analysis.
3. The hierarchy is one-directional and it is difficult to accommodate feedback relationships between lower hierarchies and higher hierarchies[5].
4. The values are highly aggregated and difficult to reflect the degree of uncertainty in the estimation of the values.
5. It is also difficult to develop the relationships between alternatives and values in resource allocation applications where the measure of a value changes with the amount of resources allocated as in the case that profitability increases usually nonlinearly with the amount of investment.

2.3. Utility Theory Approach

Utility theory was first formalized in modern context by John von Neumann and Oskar Morgenstern in their 1944 classic, "Theory of Games and Economic

---

[5] Professor Saaty has recently developed the concepts of Analytic Network Process to remedy this deficiency.

Behavior." The theory uses a set of axioms as the mathematical basis for quantifying the degree of relative preference as the utility of a particular value for the decision maker. These axioms, or commonly accepted truths, are stated in a simplified version below:

Let U(i) be the utility or the degree of relative preference by the decision maker for Value i among a set of competing values.

Axiom (1) Completeness and Rankability:

For values i and j, a decision maker can have only one of the following three preferences:

  (i)   Value i is preferred to Value j, then U(i)>U(j)
  (ii) Value j is preferred to Value i, then U(i)<U(j)
  (iii) Value i is equally preferred to Value j, then U(i)=U(j)

Axiom (2) Transitivity and Consistency:

If U(i)>U(j) and U(j)>U(k), then U(i)>U(k).

Comments: Although this axiom seems intuitive, it is not unusual that some decision makers display cyclical preferences, i.e., Value i is preferred to Value j, and Value j is preferred to Value k, but Value k is preferred to Value i. This generally occurs when the decision maker has considerable ambivalence about the values and/or when the preferences are made over different periods of time. However, for a rational decision maker, this axiom should hold for preferences made *simultaneously in time*.

Axiom (3) Substitutability:

For Values i and j, if U(i) = U(j), then these values are totally substitutable for each other.

Comments: Some decision makers may have difficulty with this axiom, especially when the two values are very different in nature, such as a tangible monetary value and an intangible environmental value. The difficulty arises generally because of different implicit assumptions made in the comparisons. For example, the utilities of a ton of greenhouse gas reduction and of $50,000 to the decision maker may be the same under one set of assumptions but different under another set of assumptions. If the utilities are indeed the same under all assumptions, then these values should be truly substitutable.

Axiom (4) Computability of Expected Utility:

If a lottery L has two possible outcomes, e.g., one has probability p of achieving Value i and the other has probability 1–p achieving Value k, then the utility of the lottery is defined and computed by the expected value U(L) = pU(i) +(1–p)U(k).

This axiom can be generalized to n possible outcomes, with probability p_i of achieving Value i and all p_i's summing to 1. Then $U(L)=\Sigma_i\, p_i U(i)$.

Axiom (5) Continuity of Expected Utility:

For Values i, j, and k, if $U(i)>U(j)>U(k)$, then there exists a probability p^* that $U(j) = p^*U(i) + (1-p^*)U(k)$. In this case, the lottery with probability p^* of yielding Value i and probability $1-p^*$ of yielding Value k is the *indifference lottery* to Value j and Value is the *certainty equivalent* of the indifference lottery.

With these axioms, a decision maker can now quantify the relative preferences of different values. Specifically, the decision maker will first identify two extreme values i and k, one the decision maker prefers the most, such as a huge monetary gain, or even an intangible value like great ecstasy, and the other the decision maker prefers the least, such as a monetary loss, or an intangible value like extreme agony. The decision maker artificially sets a very high utility for the most preferred Value i and a very low utility for the least preferred Value k. Then these axioms can be applied to quantify all other values in between the two extremes through a lottery scheme as illustrated by the following examples[6].

Example 1: Utility of Monetary Value

To determine the utility of a monetary value, say $500,000, between the two extremes of $0 and $1 million, the decision maker can assign U($1 million)=100 and U($0)=0. Then the decision maker is asked to choose between two alternatives: (1) $500,000 for sure and (2) a two-outcome lottery with probability p yielding $1 million and probability 1–p yielding $0.

Assume that the utility of money is a non-decreasing function of the amount of money; i.e., the decision maker will not prefer less money to more money. By Axioms (1) and (2), clearly, if p=0, then the decision maker will choose the alternative of $500,000 for sure. On the other hand, if p=1, then the decision maker will certainly choose the lottery. By Axiom (5), there exists a probability p^* such that the decision maker will be indifferent between $500,000 for sure and the lottery with probability p^* yielding $1 million and probability $1-p^*$ yielding $0. By Axioms (3) and (4), U($500,000)=$p^*$U($1 million)+$(1-p^*)$U($0) =$100p^*$.

Example 2: Utility of an Intangible Value

To determine the utility of an intangible value, such as the Prestige of national recognition, in relation to the utility of net monetary profit, the decision maker needs to identify a high monetary value, say $100 million in net profit, that is more

---

[6] Like temperature expressed in terms of degree of Fahrenheit or Celsius and like altitude expressed in terms of foot or meter, utility is a measure of value in terms of the degree of relative preference by the decision maker. Thus, the utility of a particular value can change with the utilities assigned to the extreme values.

preferred to this intangible Prestige value, and a monetary value, say $0 in net profit, that is less preferred to this intangible value. Again by Axiom (5), there exists a probability p* for which the decision maker will become indifferent between such a Prestige value and a lottery that yields $100 million net profit with probability p* and $0 net profit with probability 1–p*, and by Axiom (4), U(Prestige) = 100p*.

On the other hand, assuming that a decision maker prefers the value of Profitability to the value of Quality and the value of Quality to the value of Prestige, the decision maker can then estimate the utility for Quality in terms of the utilities of Profitability and Prestige. Specifically, the decision maker could assign a high utility say 50 for Profitability and a low utility say 10 for Prestige. By Axiom (5), there exists a probability p* for which the decision maker will be indifferent between having the Quality value alone for sure and a lottery that yields Profitability value alone with profitability p* and Prestige value alone with probability 1–p*, and by Axiom (4), U(Quality) = 50p* + 10(1–p*) = 40p*+10.

Note that if the decision maker assigns different utilities to Profitability and Prestige, even though the probability p* remains the same for the indifference lottery, as a relative measure, the utility for Quality may be different. Specifically, if U(Profitability)=60 and U(Prestige)=5, then by Axiom (4), U(Quality) = 60p* + 5(1–p*) = 55p*+5.

Example 3: Utility of a Candidate Technology

As an approach to measure the decision maker's degree of preference, utility theory can also be applied to objects other than value. For example, it can be used to determine directly the degree of preference for a candidate technology B by the decision maker. In this case, the decision maker needs to identify a technology A that is more preferable to B and set artificially U(A) to say 100, and a technology C that is less preferable to B and set artificially U(C) to say 0. Then by Axiom (5), there exists a probability p* for which the decision maker will feel indifferent between technology B and a lottery that yields technology A with probability p* and technology C with probability (1–p*), and by Axiom (4), U(B)=100p*. Again, if different utilities were assigned to technologies A and C, then the utility of technology B may vary even though the indifference probability p* remains the same.

2.4. Risk Attitude and Risk Premium

When utility theory is applied to monetary value, a decision maker generally will have utilities for money that are not proportional to the quantities of money. This non-proportionality reflects the decision maker's risk attitude towards money.

If a decision maker is *risk-avoiding*, he or she would prefer a small amount of sure money to a lottery that has a higher expected monetary value but carries a significant risk of yielding an outcome with little monetary value or even possibly a monetary loss.

As a specific example, such a risk-avoiding decision maker would be one who prefers $500,000 to a lottery that has a 60% probability of yielding $1 million and

40% probability of yielding $0, even though the lottery has an expected monetary value of (0.6)($1 million)+(0.4)($0)= $600,000, which is higher than $500,000. If this risk-avoiding decision maker sets U($1 million)=100, and U($0)=0, and feels indifferent between getting $500,000 sure money and a lottery that has a probability of 0.7 in yielding $1 million and 0.3 in yielding $0, then to this decision maker, U($500,000) = (0.7)(100)+(0.3)(0) = 70.

For such a risk-avoiding decision maker, the difference between the expected monetary value of the lottery and the sure amount of money that is equivalent in utility to the lottery is called the *risk premium*, i.e., it is the amount of expected monetary gain the decision maker is willing to give up in order to avoid the risk of getting the undesirable outcome from choosing the lottery. In the above example, the risk premium for this risk-avoiding decision maker is [(0.7)($1 million)+(0.3) ($0)]-$500,000=$200,000.

In insurance, the risk premium is the extra amount of money that the insurance company charges an individual customer above its expected loss for the average customer in the same class of insurance risk. For example, a technology developer carries a $10 million liability insurance policy for an annual insurance premium of $2,000. For simplicity, assume that in a given year the insurance company has a 1/10000 probability of paying out $10 million for liability damage claims for the developer and a 9999/10000 probability of paying out nothing. Then the risk premium for the technology developer is $2,000 – [(1/10000)($10 million) + (9999/10000)($0)] = $1,000.

On the other hand, the decision maker is *risk preferring* if he or she prefers a lottery that has a possible outcome of high monetary value but a low overall expected monetary value to the alternative of getting a sure monetary value that is greater than the expected monetary value of the lottery. In this case, the decision maker prefers the thrill of a risky gamble for the chance of a high monetary value outcome to accepting a lower amount of sure money.

As a specific example, such a risk-preferring decision maker would be one who prefers a lottery with a 40% probability of yielding $1 million and a 60% probability of yielding $0 to getting $500,000 for sure, even though the lottery has a lower expected monetary value of $400,000. If this risk-preferring decision maker sets U($1 million)=100 and U($0)=0 and becomes indifferent between a lottery with a 35% probability of yielding $1 million and a 65% probability of yielding $0 and the alternative of getting $500,000 for sure, then U($500,000) = (0.35)(100) + (0.65)(0) = 35.

Finally, if the utility of money is proportional to the amount of money, then the decision maker is *risk neutral*. In other words, the decision maker would be indifferent between a lottery with an expected monetary value and that amount for sure. As a specific example, if this risk-neutral decision maker sets U($1 million)=100 and U($0)=0, then U($500,000)=50.

As shown in Figure 2.2, for a risk neutral decision maker, the utility function U(x) for monetary value x is a straight line with U(x)=cx, where c is a proportionality constant. On the other hand, U(x) will be a concave curve above the straight line for a risk-avoiding decision maker, and a convex curve below the straight line for a risk-preferring decision maker[7].

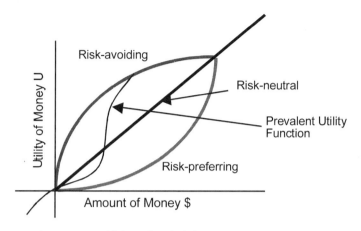

Figure 2.2. Utilities of and Risk Attitudes toward Money

The risk attitude of a decision maker generally depends on two factors: One factor is the decision maker's personality, whether the person enjoys the thrill of winning big despite unfavorable odds or is more comfortable in avoiding the risk for a possible loss even when the chance is small. The other factor is the amount of resources available to the decision maker. If there are large amounts of resources available, then most decision makers tend to be more willing to take risks and become at least risk-neutral. On the other hand, if there are little resources available, then decision makers tend to feel less willing to take the risk of a low return or a significant loss even though the probability may be small.

Because of the effect of resource availability, many decision makers would display a combination of these risk attitudes as shown by the S-shaped prevalent utility function in Figure 2.2. In this case, when the stakes are small, such as purchasing a lottery ticket that costs $1 but has an expected payoff of $0.5 in a game with 1 out of 6 million chances of winning the $3 million jackpot and $0 otherwise, many decision makers would be risk-preferring and take the gamble. On the other hand, when the stake is high, such as a $100,000 investment in a high-tech stock with a 0.2 probability of a net profit of $500,000 but a 0.8 chance of a total loss in a year, many decision makers would be risk-avoiding and prefer safe investments such as a federally insured bank savings account with a low annual return, say $5,000, which is much less than the 0.2($500,000)+0.8(-$100,000)=$20,000 expected net profit of the risky investment.

The risk attitude concept applies not only to monetary value but also to other values, such as the value of time, as illustrated by the following example. Assume

---

[7] A concave curve is defined as a curve that all points on the straight line connecting two points on the curve lie either on or below the curve, and a convex curve is defined as a curve that all points on the straight line connecting two points on the curve lie either on or above the curve.

that there are two alternative methods of developing a technology: one is a sure but slow method that takes a year to develop the technology, and the other is a risky method that has a 60% probability of developing the technology in 0.5 year and a 40% probability of developing it in 1.5 years. If the decision maker prefers the sure and slow method to the risky method, then he or she is a risk-avoider as the risky method has actually a shorter expected development time of 0.9 year than the 1-year development time of the sure and slow method. On the other hand, if the decision maker prefers a risky approach that has a 40% probability of developing the project in 0.5 year and a 60% probability of developing it in 1.5 years to the sure but slow method, then he or she is risk-preferring, as the expected development time for the risky approach is 1.1 year, which is greater than the 1-year development time of the sure but slow method.

Because risk attitude is subjective and varies among decision makers, it will be difficult to standardize or generalize in a typical technology decision making process. For simplicity of discussing the various search methods for the best alternative to maximize the utility of the monetary value, in the remainder of this book, we will assume the decision maker to be *risk-neutral*, i.e., the utility for money of the decision maker is proportional to the amount of money. Thus, maximizing the expected utility of the monetary value of the decision maker is equivalent to maximizing the expected monetary value.

Major Advantages and Disadvantages

The major advantages of the Utility Theory approach in quantifying values are in the intuitive appeal of the axioms and the ease of the utility estimation procedure.

However, the estimation procedure requires in-depth self examination, which may not appeal to some decision makers. Furthermore, it is based on two arbitrarily set extreme utilities. As a result, the intermediate utilities estimated vary with the extremes and the variation may not be internally consistent among different sets of extremes.

Specifically in Example 1 shown earlier in this chapter, the decision maker has estimated U($0.5 million) to be 100p*, based on the indifference between a lottery of probability p* for getting $1 million and probability 1–p* for getting $0 and the alternative of getting $0.5 million for sure.

Now, using a new lottery with outcomes $0 and $2 million and setting U($0)=0 and U($2 million)=200, a decision maker may be indifferent between a lottery of probability x for getting $2 million and probability 1–x for getting $0 and the alternative of getting $1 million for sure. Furthermore, the decision maker may be indifferent between a lottery of probability y for getting $2 million and probability 1–y for getting $0 and the alternative of getting $0.5 million for sure. Thus, for the decision maker using the new lottery, U($1 million)=200x and U($0.5 million)=200y. In this case, if the decision maker is totally consistent, then y should be equal to xp*, as indicated in Example 1. However, because of fluctuations in human perceptions and judgments of different extremes, y is often observed as not equal to xp*, and hence the utility estimates can be unstable.

Moreover, in some cases, actual human preferences may not follow the axioms of utility theory. One of the most famous observations of such violation is the Allais Paradox, which arose when people were surveyed about the following choices:

A. Choice between

A1: $1 million cash

A2: A lottery with a 10% chance of winning $5 million, an 89% chance of winning $1 million, and a 1% chance of winning $0.

B. Choice between

B1: A lottery with a 11% chance of winning $1 million and an 89% chance of winning $0

B2: A lottery with 10% chance of winning $5 million and 90% chance of winning $0.

C. Choice between

C1: $1 million cash

C2: A lottery with a 10/11 chance of winning $5 million and 1/11 chance of winning $0.

By the axioms, if the decision maker is risk-neutral, then the second alternative of each choice would be preferred because it yields a higher expected monetary value. On the other hand, if the decision maker is risk-avoiding, then by the axioms, to be consistent, the first alternative of each choice should be preferred. However, French scientist Maurice Allais observed that most people strongly preferred A1 to A2, and C1 to C2, but B2 to B1, which would be a contradiction of the axioms. This and other paradoxes may indicate that either the axioms are not complete in their description of the logic of human preferences or human perceptions may have produced distortions about the chances and payoffs of various alternatives that cause violations of these axioms. In any case, caution should be exercised in the application of utility theory to quantify values.

2.5. Equivalence and Reconciliation between the Analytic Hierarchy Process and the Utility Theory Approach

Both AHP and Utility Theory quantify values by measuring the degrees of relative preferences of these values to the decision maker. Thus, they are theoretically equivalent.

Specifically for the example in AHP, Profitability has the highest weight of 0.71 and Prestige has the lowest weight of 0.07. Since utility is a measure of the degree of relative preference, we can set U(Profitability)=71 and U(Prestige)=7. Then, theoretically U(Quality) should be 22. In other words, the decision maker should be indifferent between Quality for sure and a lottery with probability $p^*=(22-7)/(71-7)=0.234$ of yielding Profitability and $(1-p^*)=0.766$ of yielding Prestige. However, in reality, a decision maker would often be unable to produce a probability p^* for the indifference lottery close to this totally consistent ideal value of 0.234.

In general, because AHP has the more rigorous basis of hierarchical structuring, pair-wise comparisons, and consistency check, the quantification of value process tends to be more reliable than that based on the Utility Theory. Thus, in the above example, if the probability p^* determined by the indifference lottery is 0.3, then

U(Quality) = (0.3)(71) + (0.7)(7) = 26.2. But this U(Quality) can be reconciled with that obtained from AHP by adjusting p* to 0.234, which is always possible because 22 lies between 71 and 7 and p* is simply the interpolation ratio (22–7)/(71–7) obtained by solving the equation p*(71)+(1–p*)(7)=22.

However, if the two methods produce significantly different utilities for a value lying between identical extreme values, then it will be useful to re-examine the detailed implementations of both methods to be sure that they indeed reflect consistent judgments by the decision maker. Again, using the above example, if the decision maker applies the utility theory and becomes indifferent between Quality for sure and a lottery with a 70% probability of yielding Profitability and a 30% probability of yielding Prestige, then by Axioms (4) and (5) of utility theory, U(Quality) = (0.7)(71)+(0.3)(7) = 51.8, which is significantly different from the utility of 22 obtained from AHP. Although AHP is theoretically more reliable, it will still be useful to recheck the pair-wise comparison matrix of AHP as well as the estimation process based on utility theory to uncover the root cause of this significant differences in the utilities for Quality obtained by these two methods. In the end, with such large difference, it is possible that both estimations would require some adjustments to reconcile the utilities for Quality to maintain the consistency in the decision maker's judgments about the degree of relative preference of these values.

2.6. References

Analytic Hierarchy Process

- Saaty, T., *The Analytic Network Process: Decision Making with Dependence and Feedback,* RWS Publications, revised edition 2001.
- Saaty, T., *Decision Making for Leaders,* RWS Publications, 1999/2000 revised edition.
- Saaty, T., *Multicriteria Decision Making: The Analytic Hierarchy Process,* RWS Publications, extended edition 1990.
- Software: Expert Choice, www.expertchoice.com

Utility Theory

- Barbara, S., Hammond, P., and Seidl, C. (Ed.), *Handbook of Utility Theory*, Springer, 1999
- Fishburn, P., *The Foundations of Expected Utility*, Springer, 1982
- Neumann, J. and Morgenstern, O., *Theory of Games and Economic Behavior*, Princeton University Press. 1944 second edition. 1947
- Software: DecisionPro, Vanguard Software Corporation, www.vanguarddsw.com

2.7. Exercises

Problem 2.1

For the computer purchase exercise in Problem 1.4, apply both the Simultaneous Rating approach and the Analytic Hierarchy Process to the three major values: Affordability, Quality, and Performance, and compare the results for your relative preferences of these values.

Problem 2.2

For four values A,B,C, and D, a decision maker has the following pair-wise comparison matrix and rankings of the confidence in the comparisons:

	A	B	C	D
A	1	3	1/3	2
B		1	2	1/2
C			1	4
D				1

(1) B vs. C Highest
(2) B vs. D
(3) C vs. D
(4) A vs. B
(5) A vs. C
(6) A vs. D Lowest

Use the ranking to make the appropriate modifications of the pair-wise comparisons so that the matrix would become totally consistent.

Problem 2.3

Based on the application of the Analytic Hierarchy Process (AHP) in Problem 2.1, let V1 be your weight or rating for the most preferred value and V3 be your weight or rating for the least preferred value. Set the utilities of the most and least preferred values respectively as U(most preferred value)=V1 and U(least preferred value)=V3. Apply Axioms (4) and (5) of the Utility Theory and determine your subjective judgment for the probability p that will make a lottery of getting V1 with probability p and V3 with probability 1–p indifferent from the alternative of getting the value in between for sure. Then estimate the utility of this value between the most and least preferred values. Is this estimated utility equal to the weight or rating for the value in between obtained by AHP? If not, in which results are you more confident? Furthermore, if the two results are different, how can you reconcile them?

Problem 2.4

The CEO of a technology company is comparing two alternatives:

(a) improving an existing technology, that is *certain* to produce a moderate profit
(b) developing a new technology with the following possible outcomes:

(1) Successful development with major potential profit as well as *enhancing company image and future markets*, which would be most preferable for the CEO

(2) So-so development with moderate potential profit

(3) Unsuccessful development resulting in a significant financial loss, which would be least preferable for the CEO

The CEO believes the probabilities of outcomes (b1), (b2), and (b3) to be 25%, 50%, and 25% respectively.

The CEO prefers Outcome (b1) the most and Outcome (b3) the least. Furthermore, after much contemplation, the CEO feels that she would be indifferent between the outcome of Alternative (a) and a lottery with an 80% chance of yielding Outcome (b1) and a 20% chance of yielding Outcome (b3).

Furthermore, she would feel indifferent between Outcome (b2) and a lottery with a 70% chance of yielding Outcome (b1) and a 30% chance of yielding Outcome (b3).

The CEO assigns a utility of 100 to Outcome (b1) and −50 to Outcome (b3). Use the utility theory to estimate the utilities of Outcomes (a) and (b2).

Problem 2.5

For the ease of analysis, utility functions are often stylized into simple mathematical forms. For each of the following four utility functions of monetary value x for a decision maker, assume that the U($1 million)=100, and determine the utility of $500,000 and whether the decision maker is risk preferring or avoiding. Justify your answer.

(a) $U(x) = x^2$,

(b) $U(x) = x^3$,

(c) $U(x) = x^{1/2}$

(d) $U(x) = x^{1/3}$

3 Create Innovative Portfolio Alternatives

The effectiveness of technology portfolio planning and management depends heavily on the quantity and quality of the alternatives the decision maker will consider. If either too few alternatives are considered or many alternatives are of poor quality, the decision maker will not be able to find the alternative which will provide the optimal value to portfolio planning and management, and valuable resources may be misspent sifting through the many irrelevant or inferior alternatives.

Generating high-quality alternatives productively requires creativity. At the same time, to accurately model the relationships between alternatives and values as well as to efficiently search for the optimal choice among a large number of available alternatives also requires creativity. Unfortunately, there is no known foolproof systematic approach for developing creativity. This chapter will present three of the most widely used approaches to stimulate creativity in the thought process: Productive Brainstorming, Lateral Thinking, and the Theory of Inventive Problem Solving.

3.1. Productive Brainstorming

Brainstorming is a popular group process for stimulating and eliciting innovative thoughts. In the process, the participants are asked to express their views for either creating alternatives or suggesting a search method for the optimal choice following established ground rules:

1. Relaxed atmosphere

   Since creativity requires a relaxed mind free of rigid structures and restrictions, the brainstorming process needs to be conducted in a relaxed atmosphere with a pleasant physical environment, informal attire, and friendly interactions. A skillful facilitator is often used to encourage expressions and interactions by making the participants feel relaxed and comfortable.

2. No criticism

   The purpose of brainstorming is to induce innovative thinking. Some of the results may seem outlandish or even bizarre at first glance. There must be no criticism of any views expressed by a participant during the process no matter how strange or impractical they may appear.

3.   Piggybacking encouraged

To stimulate creative thinking, participants in the process are strongly encouraged to piggyback, i.e., expand or extend on the thoughts expressed by fellow participants.

The thoughts and suggestions expressed by all participants are recorded in real time, and then sorted, and evaluated for their practicality for implementation.

Because of its simplicity and general effectiveness, brainstorming has gained such popularity for stimulating creative thinking that it is now part of the common English vocabulary. It is also an effective process for generating diverse ideas about how to solve a problem, or different perspectives about possible future changes. Thus, it is used extensively and repeatedly in the scenarios analysis process to be discussed in Chapter 5 and in the project portfolio monitoring system to be discussed in Chapter 8.

However, the group process for brainstorming needs to be conducted productively so that a wide range of innovative thoughts can be stimulated while time will not be wasted on frivolous ideas and discussions. Traditionally productive brainstorming relies on an experienced and skillful facilitator who will keep the discussion on focus to ensure that it will not go off on a tangent.

Now, with the advances in information technology and artificial intelligence, computers can be used to further increase the productivity of the brainstorming process by eliciting inputs through both on-line and off-line interactions among participants, and by assisting in the compilation and integration of diverse views and ideas through statistical pattern recognition and correlation analysis software tools[8].

3.2. Lateral Thinking

Lateral thinking was created by British innovator Edward de Bono in 1967, as a contrast to conventional logical or linear thinking. Specifically, de Bono argued that in conventional linear thinking the thought process follows a logical direction developed from the assumptions of particular perception of the reality. Lateral thinking promotes changes in the perceptions and assumptions to create new directions or alternatives.

The following is a simple example of the application of lateral thinking to a technology portfolio decision.

The senior executives of a technology company are deadlocked in a debate about the company's choice of allocating its limited resources to one of two competing technology development alternatives:

1.   developing a new technology with potentially high returns but also high risks;

---

[8] One such software is the Angler[TM] program developed by the Artificial Intelligence Center at SRI International.

2.   improving an existing technology with low risks but also relatively low returns.

A consultant is retained to resolve the impasse. Based on the principle of lateral thinking, the consultant points out that the choice does not have to be limited to either (1) or (2), but could include many other, potentially even better, alternatives that may have higher returns with lower risks, such as:

1.   different timing: improve the existing technology for the present and use the returns to develop the new technology in the next period;
2.   different level: take a small amount of resources to conduct a pilot project on the new technology to further understand and find ways to reduce the risk, while spending the bulk of the resources to improve the existing technology without significant reduction in return;
3.   different organization: take a moderate amount of resources to develop a joint venture with another organization to reduce the risk of developing the new technology while spending the remaining resources to improve the existing technology with moderate reduction in return;
4.   different development: use part of the resources to purchase the license of a similar new technology that has already been developed to significantly reduce its developmental risk and concentrate on achieving high returns.

The list can go on. The principle here is to challenge the perception and assumption that the company is limited to two choices: either (1) develop the new technology or (2) do the seemingly logical opposite of improving the existing technology. Through the lateral thinking process, instead of being forced to choose one of the two opposing alternatives, the decision maker may consider many other innovative alternatives.

By focusing on the perceptions and assumptions, lateral thinking can also help organize the often random process of brainstorming and provide a more productive approach to creative thinking.

3.3. Theory of Inventive Problem Solving (TRIZ)

TRIZ, the Russian acronym for the Theory of Inventive Problem Solving, was invented in the 1960s by Russian patent engineer Genrich Altshuller, who concluded through the analysis of thousands of patents that there are scientific approaches for developing creativity. Altshuller summarized these approaches into 40 inventive principles, which have since been applied worldwide by numerous large corporations. Although the principles were initially developed mostly for engineering products, they have since been applied to many other fields including business operations, architectural designs, food technologies, and social programs.

A full discussion of TRIZ is beyond the scope of this chapter. Interested readers can find additional information from the references listed at the end of the chapter. A few major principles synthesized from Altshuller's work and some simple examples of their applications are given below.

Principle of contradiction

Altshuller recognized that the need for invention is motivated by a contradiction or conflict among the values in an existing situation. As illustrated by the example in Section 3.2, the fundamental contradiction in the technology development decision is between the desire for high return and the desire for low risk. Traditional approaches often would suggest either a trade-off among these values to reach a middle ground solution or a request for additional resources to implement competing alternatives. Altshuller argues that such approaches are not innovative at all, but merely compromises or ineffective allocation of valuable resources; instead invention should focus on finding ways to dissolve the contradiction through a creative framework of thinking that results in alternatives with not only high return but also low risk.

The common contradiction between quality and cost of product manufacturing is another example. Traditional approaches would often suggest strengthening the quality control process, which may result in a large number of product rejects and increase the eventual cost of manufacturing. TRIZ thinking would suggest examining the root causes of poor quality production and redesigning the manufacturing process for quality, which would result in high quality products and low overall cost.

Principle of ideality

An ideality or ideal solution to a problem would be dissolving the contradiction with only available resources at hand. Although it may not always be possible, such an ideality forces the decision maker to think creatively rather than resort to easy compromise or request more resources. Alternatives (3)-(6) suggested for the example in Section 3.2 are applications of this principle.

A company developing an automated metal cutting and machining process is a manufacturing example. Test operation of the automated process found that metal chips repeatedly jammed the cutting machine and rendered it inoperable. These metal chips had previously been swept away by the operator in the manual process. Traditional thinking would suggest emulating the manual process by adding an automatic broom or blower to the cutting machine to remove the metal chips. These traditional solutions would generate new complexities in operation and maintenance. The Principle of Ideality instead suggests that the ideal solution should be to dissolve the contradiction between the desire for a simple automation process and the desire for trouble-free operation by using available resources at hand without resorting to adding more parts requiring additional resources and increasing complexity. Based on this principle, a solution would be turning the machine upside down to use an existing resource, gravity, to remove metal chips automatically.

Principle of root cause analysis

Very often the root causes of a contradiction are not apparent. The decision maker needs to search for these root causes by systematically examining and tracing the source of the contradiction. For the example in Section 3.2, the root cause of the contradiction is not in the apparent conflict between alternatives (1) and (2), but between the desire for high return and the desire for low risk. Thus, a creative solution would be to dissolve this contradiction by finding a way to increase return without increasing risk or reduce risk without reducing return.

Similarly, for the metal cutting machine example cited earlier, the root cause for the contradiction is not between the benefit and cost of an automated process, but between the desire for a simple process which had resulted in the metal chips jamming the machine and the desire for a trouble free operation. Thus, a creative solution would be to dissolve this contradiction by finding a way for the chips to fall away from the machine without sacrificing the simplicity of the process.

Principle of related solutions

Many inventions can be applied to different fields. This principle helps categorize inventions in different fields so that they can be easily searched, retrieved, and used for similar applications in other fields. As an example, Boeing Company has developed a technology for lifting seats into an aircraft assembly by adopting and modifying a hay loader used in the 18th century. Similarly, application of the financial option theory to technology and other real property portfolios has resulted in the Real Options theory.

Principle of unrelated combinations

This principle stimulates creative thinking by finding ways to combine apparently unrelated characteristics. Altshuller gave an example of combining the characteristics of a leopard and a pencil. The creative thoughts of combining these characteristics may result in pencils with leopard spots or leopard-like pencils. Like the Principle of Ideality, this principle forces the decision maker to think outside the box.

Principle of technology evolution

Altshuller also observed that many technologies evolve over time following similar and often predictable patterns. One line of evolution he postulated is the sequential and/or parallel transition from mechanical to electrical to chemical to electronic systems. As an example, the table below indicates two diverse unrelated technologies following this line of evolution:

Technology Evolution	Communication	Teeth-Cleaning
Mechanical	Manual delivery	Tooth pick, manual tooth brush
Electrical	Telegraph, telephone	Electric tooth brush
Chemical/electrochemical	Photocopying, telefaxing	Dentifrice, mouth wash for plaque removal
Electronic	Wireless	Potential use of electronics to remove plaque

This principle provides a tool for forecasting and providing insights into future technology development.

3.4. References

General

Gardner, M., *Aha! Insight*, W.H. Freeman, 1978

Poundstone, W., *How Would You Move Mount Fuji?*. Little, Brown, 2003

Saaty, T., *Creative Thinking, Problem Solving, and Decision Making*, RWS Publishing, 2006

Thoms, N., (Ed.) *Adair on Creativity and Innovation*, Thorogood Publishing, 2004

Von Oech, R., *A Whack on the Side of the Head*, Warner Books, 1990

Brainstorming

Rodriguez, A. and Boyce, T., Lowrance, J., and Yeh, E. "Angler: Collaboratively Expanding Your Cognitive Horizon," *International Conference on Intelligence Analysis Proceedings*, MITRE, May 2005

Lateral Thinking

de Bono, E., *de Bono's Thinking Courset*, Facts-on-File, Revised Edition, 1994.

de Bono, E., *Lateral Thinking for Management*, McGraw-Hill, 1971.

de Bono, E., *New Think: The Use of Lateral Thinking in the Generation of New Ideas*, Basic Books 1969.

TRIZ

Altshuller, G., *The Art of Inventing (And Suddenly the Inventor Appeared)*. Translated by L. Shulyak. Worcester, MA: Technical Innovation Center, 1994.

Altshuller, G., *Creativity as an Exact Science*. Translated by A. Williams, Gordon and Breach Science Publishers, 1984

Savransky, S., *Engineering of Creativity: Introduction to TRIZ Methodology of Inventive Problem Solving*, CRC Press, 2000

Rantanen, K. & E. Domb, *Simplified TRIZ : New Problem-Solving Applications For Engineers & Manufacturing Professionals*, St. Lucie Press, 2002

3.5. Exercises

For Problems 3.1–3.7, find a solution for each and comment on the creativeness of your solution.

Problem 3.1

You are driving a two-seater sports car and come to a bus stop where three persons waiting for the bus are in need of a ride:

- A sick old lady who direly needs to be transported to the emergency room of the nearest hospital
- One of your best friends who has done you a big favor in the past and now urgently needs to get to an important meeting downtown
- A person of your dreams whom you have been dying to meet and may never have another opportunity if you don't do it now

How should you decide which person you would give a ride?

Problem 3.2

You are asked to use a barometer and a rope to measure the height of a tall building to which you have total access.

Problem 3.3

A consumer product company sells a small, very flexible product, packaged in quantities of 10 each. An outside supplier manufactures the product and bundles it into tens with rubber bands. The consumer products company unbundles and repackages it for final sale. The unbundling and repackaging process has caused many ergonomic injuries and additional costs, while the final packaging is still hard to use for the customers. What should the company do?

Problem 3.4

There are 12 balls identical in appearance but one of them is either heavier or lighter than the others. Use a scale three times to find the odd ball and also know whether it is heavier or lighter.

Problem 3.5

The president of a country was hosting a state banquet for a visiting head of state, where many of the national treasures were on display. At the banquet, a security staff member reported to the Director of Security that the wife of the visiting head of state was seen to have sneaked an extremely valuable piece of the national treasure into her purse. As the Director of Security, how would you be able to retrieve this valuable piece of national treasure without creating an international incident?

Problem 3.6

The following is based on a true story: Susan has recently immigrated to the United States with her husband, who has just started a successful career in his adopted country. As the only child, Susan received an urgent phone call about her widowed mother developing serious medical problems that require long-term comprehensive care. Susan desperately wants to provide personal care to her beloved mother.

However, visa restrictions have made it impossible for her mother to come to the U.S. On the other hand, Susan's husband has made it absolutely clear that he will not abandon his budding career to move back with Susan. Susan believes that if she goes back alone for an extended period of time, the marriage is likely to deteriorate, which she definitely wants to avoid. Susan is torn between the alternatives of caring for her mother and staying with her husband. How would you advise and help her to resolve this dilemma? Please also comment on the general characteristics of the problem and whether they often occur in many other situations.

Problem 3.7

An overseas automobile manufacturer uses inexpensive local labor to produce high quality cars at low cost. It would like to sell the cars in the U.S. market, which requires 30 days of shipping time through ocean freight. The manufacturer has considered the alternative of setting up an assembly plant in the U.S. but has found the U.S. labor cost too high to sustain the competitive edge. Suggest an innovative solution to this problem.

Problem 3.8

Write a story involving the creative solution of a problem. The story does not need to original. It can be a story you have heard from someone, a clever joke, or an interesting puzzle, but you need to indicate its creativeness. A contest will be held in class and the students will vote for the most creative story.

4 Determine and Forecast Relationships: Major Analytic Methods

Once the decision maker has clarified and assessed the values of technology portfolio selection and identified major innovative and relevant alternatives for the portfolio, he or she needs to determine the relationships between each resource allocation alternative and its impact on the values, as well as to forecast future changes in the relationships. To meet this need, ideally a set of comprehensive business, economic, environmental, and social-political models of these relationships should be developed, for which technology is only one major factor. However, such development is well beyond the scope of this book, and we will concentrate instead on:

- developing simplified near-term relationships between each resource allocation alternative for technology development or application and its *direct* impacts on the *immediate* values, such as profit, productivity, risks, local environmental effects, company reputation, and others
- forecasting future changes in the relationships through the use of intermediate relationships, such as technology capability, demand growth and market penetration of each technology portfolio alternative, which can then be used as the basis for estimating the overall business, economic, environmental, and socio-political impacts of each alternative

4.1. Develop Simplified Near-Term Relationships

There are three major approaches to developing simplified near-term relationships between resource allocation alternatives and its direct impacts on the immediate values:

4.1.1. Aggregate approach

In this approach, the impacts on values are estimated either qualitatively or quantitatively on an aggregated or 0–1 basis, in that the impact occurs only when the full amount of resource is applied. Estimates can be developed through either expert judgments or simplified analysis. The following are some typical examples:

- A $100 million investment in a solar energy technology development in the next 5 years will provide a return of $120 million in net present value and reduce greenhouse gas emissions by 1% per year.

- A $10 million investment in implementing a portfolio of information technologies in the next 3 years will increase the company's productivity by 10% per year.
- A $5 million investment in leading-edge technology development will significantly enhance the company's reputation in the technical community.

The main advantage of this approach is in its simplicity. Furthermore, the 0–1 characteristic may represent the minimum amount of resource allocation required to produce an impact.

However, this approach has the disadvantage of being overly simplistic. Its 0–1 all or nothing lumpiness characteristic may also be unrealistic and misleading.

This approach is generally used to provide a simple estimate or a first approximation of the impact on the values.

4.1.2. Proportionality approach

This approach is a modification of the aggregate approach, with the impacts on the value being approximated as proportional to the amount of resources allocated. Thus, it remains simple but avoids the lumpiness of the previous approach. Furthermore, if the proportionality characteristic is realistic, this approach can provide the basis for linear programming analysis to be discussed in Chapter 6.

However, this approach cannot impose the often necessary requirement of a minimum threshold of resource allocation before any impact will be realized from technology development or application. Moreover, it would be unrealistic to expect the magnitude of an impact to increase proportionally with resource allocation without bound.

This approach is again largely used to provide a simple estimate of the impact of alternative resource allocation on the values.

4.1.3. S-shaped curve approach

This approach estimates the impacts on values as S-shaped curves. Specifically, it indicates that the impact will be low for small amounts of resource allocation, and increases significantly when the resources allocated exceed a threshold amount. As resource allocation continues to increase, the impact will gradually reach an upper limit.

The S-shaped curve may be derived from detailed engineering-economic analysis or approximated by stylized mathematical functions. The following are two typical S-shaped mathematical functions:

(1) $V(x)$ = Value at x amount of resource allocation
           $= V_0/(1 + ae^{-bx})$

where V_0 = assumed upper limit of the impact
        a = location coefficient
        b = shape coefficient

As an example based on this functional form with $V_0 = \$100$ million, a = 1,000, and b = 1 per $million, the value, $V(x)$, of allocating x million dollars to a technology is shown in Figure 4.1.

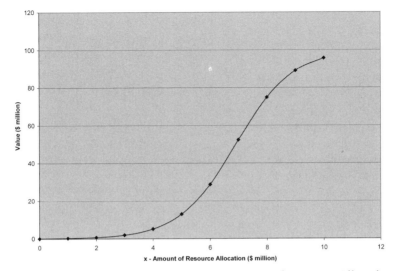

Figure 4.1. Typical S-Shaped Value as a Function of Resource Allocation

(2) $V(x) = Vo[1-e^{-b(x-xo)}]$ for x>xo
                    $= 0$ otherwise

where Vo = assumed upper limit of the impact
       b = shape coefficient
       xo = resource allocation threshold

As an example based on this functional form with Vo = $40 million, b = 0.6 per $million, and xo = $3 million, the value, V(x) of allocating x million dollars to a technology is shown in Figure 4.2.

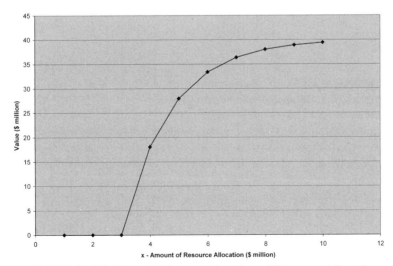

Figure 4.2. Typical S-Shaped Value as a Function of Resource Allocation with Minimum Threshold

This approach is realistic as it conforms with empirical observations of the impacts of resource allocation on values.

However, to be accurate, especially for impacts on productivity, environmental effects, and other less tangible values, this approach will require extensive modeling and analysis efforts.

This approach is generally used for detailed analysis of impacts on more tangible values, such as profit and return on investment.

4.2. Forecast Future Technology Capability, Demand Growth, and Market Penetration

Forecasting changes even in the intermediate relationships between resource allocation to alternative technologies and the future technology capability, demand growth, and market penetration is still difficult because of a lack of understanding and information. In fact, most analytic methods for forecasting these intermediate relationships are based on one of the major, largely unverified underlying assumptions presented in the following sections.

4.2.1. Underlying assumption 4A: future follows a historical pattern

The underlying assumption here is that, based on empirical observations, technology capability, demand growth, or market penetration will follow a stable and continuing historical pattern. There is generally little explanation about the logical reasons or causal mechanism of this historical pattern. However, once the functional form of the pattern is assumed and accepted by the decision maker, statistical analysis can then be applied to determine the best fit coefficients for the functional form that minimize the error between the observed data and the estimated projections. Major analytic methods based on this underlying assumption include the following:

4.2.1.1. Trend extrapolation

The assumption here is that technology demand growth will continue to follow the historical trend, which is often viewed as linear or exponential partly because they are good first approximations and partly because of their analytical simplicity. Mathematically, let Y_t be the technology demand at time t; then by this assumption,

$$Y_t = \alpha + \beta t + \varepsilon \quad \text{for a linear trend, or}$$
$$Y_t = \alpha e^{\beta t + \varepsilon} \quad \text{for an exponential trend}$$

where ε is a random error in measurements or recording, generally assumed to be normally distributed with mean 0 and standard deviation σ.

Since the exponential trend equation can be easily converted into a linear one on the logarithmic scale as $\text{Log } Y_t = \text{Log } \alpha + \beta t + \varepsilon$, we will focus on the linear trend for the remainder of the discussion in this section.

For the linear trend, α and β are generally unknown, and we will need to estimate them respectively from the observed historical trend data T_t through the following estimated relationship:

$$T_t = a + bt,$$

Specifically, linear regression analysis is used to estimate the best-fit values for a and b that minimize the sum of the squares of the errors between observed trend data T_t and estimated trend values a+bt (hence these coefficients are often called the least square best fit estimates).

Formulas for the Least Square Estimates of the Slope b and Intercept a are summarized below:

$$b = [\Sigma t T_t - (\Sigma t \ \Sigma T_t)/n]/[\Sigma t^2 - (\Sigma t)^2/n]$$
$$a = (\Sigma T_t/n) - b(\Sigma t/n)$$

where   n = number of historical time periods

Under the assumption that the random errors at different times have independent, identical normal distributions with mean 0 and standard deviation σ, the ratio, b/s, between the estimated slope b and the sample standard deviation of the statistical error, s, has a student t distribution and can be used for testing the null hypothesis, Ho: $\beta=0$. A rejection of this hypothesis would indicate the linear regression equation has a statistically significant slope. Furthermore, given the assumption that the trend is stable and continuing, confidence intervals can also be constructed for the trend projection. Detailed methodologies of these additional analyses are summarized in the Appendix.

Illustrative Examples

1. A popular example of applying this assumption to technology forecasting is the Moore's Law, which forecasts that the number of transistors on a microchip will double about every 18 months as shown in Figure 4.3. This law has held since 1970.
2. As another illustrative example, a manufacturer has collected data on the demand for a computer technology it produced for the last 5 years. The historical data, together with some computed values, are summarized in the table below. The manufacturer assumes that the trend is stable and continuing and wants to project the demand for year 6.

Year t	Demand T_t (thousands of units)	tT_t	t^2
1	11	11	1
2	14	28	4
3	20	60	9
4	26	104	16
5	34	170	25
Total 15	105	373	55

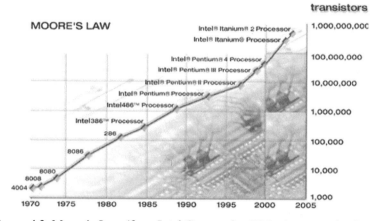

Figure 4.3. Moore's Law (from Intel Corporation Web site, www.intel.com)

Based on the assumption of stable and continuing trend, the manufacturer uses regression analysis to estimate the coefficients of the linear trend equation as follows:

$$b = [373 - (105)(15/5)]/[55 - (15)^2/5] = 5.8$$
$$a = 105/5 - 5.8(15/5) = 3.6$$

Thus, the linear trend equation is

$$T_t = 3.6 + 5.8t$$

and the demand projection for year 6 is

$$T_6 = 3.6 + 5.8(6) = 38.4$$

The observed data, the linear regression trend line, and the projections are shown in Figure 4.4.

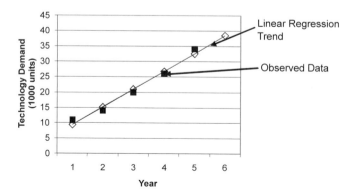

Figure 4.4. Example of Linear Trend Extrapolation

4.2.1.2. Regular fluctuations

Technology demand growth is assumed here to follow not only a stable trend but also regular fluctuations. These fluctuations are called seasonal for periods within a year, and cyclical for periods longer than a year. Because regular seasonal fluctuations are more easily observed than longer term cyclical fluctuations, most applications of this assumption concentrate on replicating seasonal fluctuations for forecasting purpose using the following multiplicative model:

$$Y_t = T_t \times S_t \times I_t$$

where Y_t = the technology demand at time t

T_t = the technology demand trend value at time t

S_t = the seasonal index at time t, which is expressed as a percentage of T_t

I_t = the irregularity index at time t, representing all unknown effects impacting on the demand as expressed as a percentage of T_t

A series of moving averages are used to reduce the seasonal and irregularity effects in order to extract the underlying trend and seasonal indexes. The computational procedure is outlined below and demonstrated through a specific numerical example that follows.

Computational Procedure

Step 0.   Collect m(>1) full years of observed demand data for the n seasons in a year

Step 1.   Calculate the Adjusted Seasonal Indexes:

1.1   Compute consecutively the n-season moving averages (n-SMAs). For m years, there will be (m–1)n + 1 such moving averages.

1.2   If n is even, compute a series of centered moving averages (CMAs) by averaging the n-SMAs of two consecutive seasons so that each CMA is centered on a particular season. If n is odd, then the n-SMAs are CMAs as they are already centered.

1.3   Divide each observed demand by the corresponding CMA to identify the seasonal-irregular effect in the data series.

1.4   For each of the n seasons, average the m–1 computed seasonal-irregular values for that season as the Seasonal Index for that season.

1.5   The Seasonal Indexes of the n seasons should sum to n; if they do not, they will be proportionally adjusted so that the Adjusted Seasonal Indexes of the n seasons will sum exactly to n. (Warning: if the sum of the Seasonal Indexes is significantly different from n, say by 20% or more, the linear trend assumption may be inappropriate and needs to be reexamined,)

Step 2.   De-seasonalize the Data Series: Divide each time series observation by the corresponding Adjusted Seasonal Index for a de-seasonalized time series.

Step 3.   Use the De-seasonalized Data Series to find the best-fit Trend Equation

Step 4.   Use the Trend Equation to project the trend value at a future season of interest and Re-seasonalize the trend value as the forecast for the future season.

Numerical Example

A manufacturer has collected quarterly demand data for a solar energy panel for the last 3 years and wants to forecast the seasonal demand in year 4. The historical demand data together with computational steps 1.1 through 1.4 are summarized in the table below.

Year	Quarter	Demand (1000 Units)	4-SMA (Step 1.1)	CMA (Step 1.2)	Seas. Irreg. Value (Demand/CMA) (Step 1.3)
1	1	3			
	2	9			
			5.00		
	3	6		5.13	1.17
			5.25		
	4	2		5.50	0.36
			5.75		
2	1	4		6.00	0.67
			6.25		
	2	11		6.38	1.72
			6.50		
	3	8		6.63	1.21
			6.75		
	4	3		7.25	0.41
			7.75		
3	1	5		8.13	0.62
			8.50		
	2	15		8.50	1.76
			8.50		
	3	11			
	4	3			

Computation of Adjusted Seasonal Indexes (Steps 1.4 and 1.5)

Quarter	Seas-Irreg Values	Seas. Index	Adjusted Seasonal Index
1	0.67, 0.62	0.65	0.655 = (0.65 x 4/3.97)
2	1.72, 1.76	1.74	1.753 ... etc.
3	1.17, 1.21	1.19	1.199
4	0.36, 0.41	0.39	0.393
		Total = 3.97	Total = 4.000

De-seasonalization (Step 2)

Year	Qtr	t	Demand	Seasonal Index	De-seasonalized Demand
1	1	1	3	0.655	4.58
	2	2	9	1.753	5.13
	3	3	6	1.199	5.00
	4	4	2	0.393	5.09
2	1	5	4	0.655	6.11
	2	6	11	1.753	6.27
	3	7	8	1.199	6.67
	4	8	3	0.393	7.63
3	1	9	5	0.655	7.63
	2	10	15	1.753	8.56
	3	11	11	1.199	9.17
	4	12	3	0.393	7.63

Estimated Trend Equation and Trend Projection based on De-seasonalized Demand (Step 3)

$$T_t = 4.066 + .3933t$$
$$T_{13} = 4.066 + .3933(13) = 9.1789$$

Seasonal Adjustments for Projections (Step 4)

t	Trend Forecast	Seasonal Index	Quarterly Forecast
13	9,179	0.655	6,012
14	9,572	1.753	16,780
15	9,966	1.199	11,949
16	10,359	0.393	4,071

The observed data, the de-seasonalized data, the trend line, and the re-seasonalized projections are summarized graphically in Figure 4.5.

4.2.1.3. Persistent technology substitution

John Fisher and Robert Pry have observed empirically that once the market share f for a new technology exceeds a certain threshold value, usually around 10%, the new technology will grow first gradually and then rapidly and it will inexorably substitute

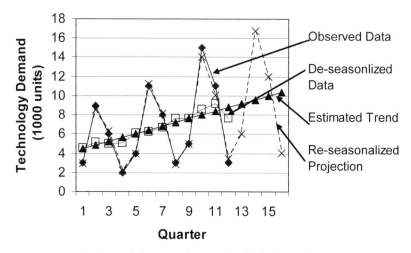

Figure 4.5. Example for Regular Fluctuations

for an existing technology until the latter is totally replaced and another new technology comes into the market. Although there is little verified understanding of the underlying reasons for this pattern, there is much empirical evidence indicating that this pattern appears to hold for technologies in many fields. Furthermore, the functional form of this assumption follows the S-shaped form, which is intuitively appealing in explaining market competition, where a strong new contender will force out an aging existing technology. As a result, the Fisher-Pry Substitution Model and other similar S-shaped models have been very popular in forecasting technology demand growth.

The general functional form under this assumption is as follows:

$$\text{Log } (f/1{-}f) = a{+}bt$$

where f is the market share of a technology, and a and b are respectively the location and shape coefficients.

Again, once this assumed functional form is accepted by the decision maker, statistical techniques can be used to find the best-fit estimates for the coefficients a and b and develop tests for the significance of this assumption.

As an example based on the Fisher-Pry model with a=−5 and b=1, the market shares of a new technology in a 10-year span are given in Figure 4.6.

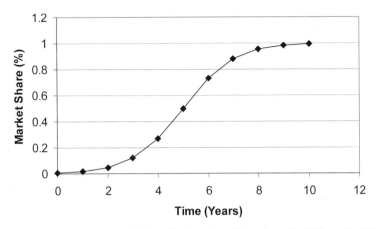

Figure 4.6. Market Share of a New Technology Based on the Fisher-Pry Model

4.2.1.4. Major Advantages, pitfalls, and applicability

The assumption that historical patterns are stable and continuing in technology demand forecasting has the major advantages of being simple, credible, and easy to use. Without another convincing alternative understanding or argument about the underlying forces for the demand growth, Occam's Razor[9] would compel us to fall back on the simple assumption that the past trend will continue into the future. This simple assumption may indeed be valid if the forecasting period is relatively short because the momentum of the underlying forces for the demand growth are likely to prevail for a short while. Furthermore, because of its simplicity and wide acceptance, many computational techniques have been well-developed to minimize errors and test statistical significance.

However, the major pitfall of this assumption is that without a deeper understanding, the continuation of historical trends and patterns, no matter how long lasting in the past, are not guaranteed in the future. As a result, when the business environment is highly volatile, as in the case of many high-tech industries, or the forecast period is relatively long, say 5 years or more, or where future changes are likely to be beyond our experience of the past, then this assumption becomes often invalid.

Since the underlying assumption is generally valid only when the momentum of the existing trend holds, it is most applicable to short-term forecasting.

4.2.2. Underlying assumption 4B: future is analogous to a well-known phenomenon

The underlying assumption here argues that it is plausible for technology demand growth to be analogous to some known biological or physical phenomena, such as

---

[9] Occam's Razor states that one should not increase, beyond what is necessary, the number of entities required to explain anything.

the organism life-cycle or the diffusion process. Practically all the technology demand growth models based on this assumption exhibit an S-shaped form.

4.2.2.1. Life-cycle model

In this case, technology demand growth is assumed to behave analogous to the life cycle of a living organism. By this analogy, once a technology survives the initial challenges in the business environment, it will start to gain vital force through user support and the demand will grow almost exponentially until organizational ossification and competitive pressures begin to slow the growth and eventually cause the decline and demise of the technology in the market. One typical life-cycle based technology demand growth model is the Pearl-Reed law given below:

$$Y_t = L/(1+ae^{-bt})$$

where Y_t = technology demand at time t
L = assumed upper limit for the demand
a = location coefficient
b = shape coefficient

Once accepted by the decision maker, statistical analysis can again be used to minimize numerical errors and produce the best-fit estimation for a and b.

As an example based on this model with L=10,000 units, a=100, and b=1 per year, the demand of a new technology in 10 years is shown in Figure 4.7.

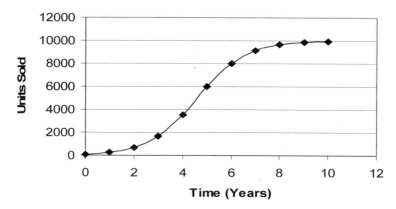

Figure 4.7. Demand of a New Technology Based on the Life Cycle Model

4.2.2.2. Physical diffusion model

In this case, technology demand growth is assumed to behave similarly to a physical diffusion process. Once the technology demand penetrates the barrier of initial resistance in the market, it will expand rapidly until reaching a saturation level that is in balance with those of other competitive technologies. A typical diffusion-based technology demand growth model is the Bass Law given below:

$$Y_t = Y_{t-1} + a(M - Y_{t-1}) + bY_{t-1}(M - Y_{t-1})/M$$

where Y_t = technology demand at time t
        a = coefficient of innovation
        b = coefficient of imitation
        M = market potential

Again, once this underlying assumption is accepted by the decision maker, statistical analysis can be used to minimize errors and determine the best-fit estimates for coefficients a and b.

As an example based on the diffusion model with a=0.01, b=0.5, and M=10,000 units, the demand of a new technology in a 15-year span is shown in Figure 4.8.

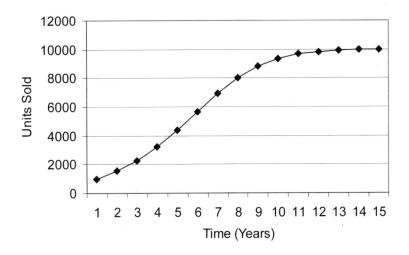

Figure 4.8. Demand of a New Technology Based on the Diffusion Model

4.2.2.3. Major advantages, pitfalls, and applicability

The major advantages of the assumption that technology demand growth is analogous to well-known biological or physical phenomena are its intuitive plausibility and theoretical elegance. Such analogies can also appear very credible.

Nevertheless, the main pitfall of this assumption is that no matter how appealing an analogy is, it may not be valid in reality.

Because this assumption provides plausible rationale about the underlying mechanism in future changes, it is most applicable to short- to medium-range forecasting.

4.2.3. Underlying assumption 4C: future can be assessed from a causal relationship-based large-scale system model

4.2.3.1. Causal relationship-based large-scale system models

Here, it is assumed that quantitative, large-scale system models can be constructed to describe the causal relationships among business competition, economic development, and even global interactions, for which technology demand growth is an integral part. Typical examples are large-scale econometric models and the System Dynamics models of J. Forrester, where a system of mathematical equations are postulated to simulate the dynamic interactions among the many factors in the economy, society, and the natural environment.

4.2.3.2. Major advantages, pitfalls, and applicability

These causal relationship-based large-scale system models are intellectually appealing and technically intricate and sophisticated. In fact, large-scale system models that can explain the causal relationships underlying the business, economic, and socio-political phenomena at a level of accuracy comparable to that which can be achieved by physical causality models for natural phenomena have been the holy grail of most forecasting professionals.

However, due to a fundamental lack of understanding about the true underlying mechanism at work in the business, economic, and socio-political environment, these causal relationship-based large-scale system models remain over-simplified or idealized theoretical constructs that can lend insights to future changes but generally cannot produce credible and verifiable quantitative forecasts.

Because of the inherent difficulty in verifying the validity of the causal relationships in these large-scale models, methods based on this underlying assumption would be most useful for research purposes and for gaining general insights about future changes.

4.3. References

Ayres, R., *Technological Forecasting and Long-Range Planning*, McGraw-Hill, 1969

Bass, F., "A New Product Growth Model for Consumer Durables," *Management Science*, Vol. 15, pp. 215–227, 1969

Fisher, J. and Pry, R., "A Simple Substitution Model of Technological Change," *Technological Forecasting and Social Change*, Vol. 3, pp. 75–88, 1971.

Forrester, J., *World Dynamics*, 2nd Edition, Wright-Allen Press, 1973

Martino, J., *Technological Forecasting for Decision Making*, McGraw-Hill, 3rd Edition, 1992

Makridakis, S., Wheelwright, S., and McGee, V., *Forecasting: Methods and Applications*, 2nd Edition, Wiley 1983

Moore, G., "Cramming More Components Onto Integrated Circuits," *Electronics*, April 19, 1965

Pearl, R. and L. Reed, "On the Rate of Growth of the Population of the United States Since 1790 and Its Mathematical Representation," in Smith D. and Keyfitz, N., *Mathematical Demography,* Springer, 1970

Porter, A., *Forecasting and Management of Technology,* Wiley, 1991

Vanston, J., *Technology Forecasting: An Aid to Effective Technology Management,* Technology Futures, 1982

4.4. Exercises

Problem 4.1

Develop an example for which the value of a technology development or application is proportional to the amount of resources allocated to the development or application.

Problem 4.2

Collect data on an actual technology development or application to estimate the relationship between resource allocation and expected value of the development or application. Examine the general shape of functional relationship.

Problem 4.3

Find an example for which the expected return on investment in a technology development or application as a function of the amount of investment is not S-shaped, and explain why it is not.

Problem 4.4

The total sales of a technology product for the 8-hour shifts in a day for an online store in the past three years are given below.

Year	8-hour shift	Ave Sales/Shift (1000$)
1	0000–0800	52
	0800–1600	224
	1600–2400	155
2	0000–0800	56
	0800–1600	244
	1600–2400	168
3	0000–0800	60
	0800–1600	268

Use these seasonal sales data to forecast the demand for the sales of the product of the 1600–2400 shift of the third year.

Problem 4.5

Collect a published technology forecast based on trend analysis that has proven to be totally inaccurate and explain the reasons for the failure.

Problem 4.6

Collect a set of actual data of market substitution between two technology products, such as VCR vs. DVD, and use the Fisher-Pry model to forecast the market shares for these products in the next 5 years.

Problem 4.7

Find an example of the demand of a technology that fits the life cycle model.

5 Determine and Forecast Relationships: Qualitative Approaches

The analytic methods discussed in Chapter 4 are appealing in their ability to produce quantitative forecasts about technology demand growth and other relationships between technology alternatives and the decision maker's values in the technology portfolio planning process. However, the underlying assumptions used for these methods generally require the relationships to be relatively stable, which may be invalid for the highly uncertain future business, economic, and socio-political environments of technology development and application. Thus, the accuracy of the quantitative forecasts can be questionable and the rigorous analytical effort required by these methods may not be justified. As a result, for a highly uncertain environment, a less rigorous but more insightful qualitative or semi-quantitative approach to understand future changes may be more appropriate, flexible, and effective. This chapter will present several major qualitative or semi-quantitative approaches to determining and forecasting relationships needed for technology portfolio planning.

Just like the analytic methods, these qualitative approaches are also based on a set of underlying assumptions. We will again categorize these approaches according to their assumptions as below.

5.1. Underlying Assumption 5A: Understanding the Future Through the Power of Expert Opinion or Collective Wisdom

Under this assumption, the decision maker believes that experts, senior executives, or the majority of technical professionals will provide the correct understanding and judgments about future changes. Major approaches with this assumption include:

5.1.1. Delphi methods

In this method, a group of experts will be first asked about their individual judgments on a particular aspect of a future technology, such as the date of availability or the level of demand in a particular year. These judgments will be collected in a blind fashion where each expert is not aware of the judgments of other experts. Then the judgments will be revealed, and experts with the most different judgments will be asked to explain the rationale of their judgments with critiques by other experts. Finally, a reconciliation between the viewpoints of these experts will be made to narrow down the differences in their judgments as much as possible.

5.1.2. Consensus approach

In this approach, a group of experts, senior executives, or experienced professionals will be asked about their opinions on a particular aspect of a future technology. Then a discussion will ensue to reach a consensus of the opinions. Different from the Delphi method, it does not reconcile the extreme differences but only tries to reach a general consensus among the participants.

5.1.3. Major advantages, pitfalls, and applicability

Approaches based on this assumption have the advantages of simplicity and relatively low cost in their implementation and the built-in credibility of expert judgments or consensus opinions.

However, the assumption has two major pitfalls:

1.   In the selection of the experts, executives, and professionals, there is a tendency to include mostly individuals with similar backgrounds or vision, which may result in a bias in the collective judgments and opinions, while ignoring the views of mavericks and outliers that could turn out to be more accurate.
2.   Even when the participants are diverse, there is no guarantee that their collective judgments and opinions are in any way valid. Bandwagon effect and herd instinct are often known to be misleading, while contrarian views may prove correct.

Because the assumption relies more on the validity of collective judgments rather than detailed reasoning, it is most applicable to forecasting technology characteristics for which there is little available historical data or widespread understanding, such as when stem cell technology will be available to cure various types of cancer or when man will land on Mars.

5.2. Underlying Assumption 5B: Understanding the Future Through the Guidance of Leading Indicators

Based on either empirical observations or logical reasoning, a set of evidence is assumed to be leading indicators for technology development or application. There is generally insufficient understanding to develop detailed analytical methods for the relationships between the occurrence of the leading indicators and the impending development or application of a technology. However, a simple threshold may be set for the leading indicators to predict the likelihood of a specific technology development or application. Major approaches based on this assumption include:

5.2.1. Patent and citation analysis

It is logical to assume that the frequency of patent filings and reference citations related to a technology would be a leading indicator for potential breakthroughs or advances in the development or application of the technology.

5.2.2. Survey of technology early adopters

In this approach, interests by early adopters identified in other technologies are surveyed and used as the leading indicator for the potential demand growth for a new technology.

5.2.3. Major advantages, pitfalls, and applicability

Approaches based on this assumption again have the advantages of plausibility and ease in implementation.

However, without deeper understanding, superficial signs and indicators could be misleading, early adopters' experience with other technologies may not be transferable, and there may be few indicators available for radically new technologies that are developed in relative isolation.

As a result, approaches based on this assumption are most applicable to short-term forecasting of future development or application of well-known technologies.

5.3. Underlying Assumption 5C: Understanding the Future Through Reasoned Analysis

In this assumption, the decision maker develops qualitative and logical reasoning-based analysis of the general structure of the business, economic, and socio-political systems for which technology development or application is an integral part. This reasoned analysis of the system structure can then be used to develop insights about the relationships between alternative technologies and the decision maker's values, as well as future changes in these relationships. Typical approaches based on this assumption include:

5.3.1. Cross-impact matrix

In this approach, the impacts of major events or trends in technology development or application on one another are qualitatively estimated and tabulated in a matrix form. Matrix analysis can then be used to estimate the combined impact of one event or trend to another event or trend.

An elementary example shown in the table below illustrates the interactions among various events that would affect demand for environmental technologies.

Table 5.1. Cross-Impact Matrix for the Demand for Environmental Technologies

	Land & soil conservation	Water system integrity	Bio-diversity	Environmental technologies
Land & soil conservation	x	(++) Reduce water contamination	(0)	(++) Demand for land manage-ment technology
Water system integrity	(++) Reduce erosion	x	(-) Water system management may restrict bio-diversity	(++) Demand for water management technology
Bio-diversity	(+) Bio-diversity can help land conservation	(+) Bio-diversity can help water system effectiveness	x	(+) Demand for wildlife management technology
Environmental technologies	(++) Restore land and soil	(++) Decontaminate water	(+) Restore wildlife	x

Where –, 0, +, and ++ indicates the direction and level of influence.

The cross-impact matrix can be made more sophisticated by quantifying the influences and even constructing time-based conditional probabilities. In the latter case, the matrix can be used for Markov process simulation for the changes in these events and trends.

5.3.2. Hierarchical influence tracing system

In this approach, factors influencing demand growth of a technology is systematically traced through a hierarchical structure. The approach first identifies the immediate primary influencing factors. It then traces changes in each primary influencing factor through a set of secondary and broader factors in the business environment. This process continues until it reaches the fundamental driving forces in the business environment. Changes in these fundamental driving forces are then estimated through a set of assumptions or theoretical models. At each level of the hierarchy, the degree of influence can be indicated by the width of the influencing arrow, and the level of knowledge about the influence can be indicated by the color or shading of the arrow.

As an example, Figure 5.1 illustrates that demand growth for a set of electricity efficiency technologies is first traced to its primary influencing factors: energy use needs, power supply availability, and electricity industry structure. Changes in these factors are in turn influenced by a set of fundamental forces: global energy situation, worldwide socio-economic developments, energy technological innovations, and government regulatory policies. Once future changes in the fundamental forces are assessed, we can systematically trace the influences and estimate the resulting demand growth for electricity efficiency technologies.

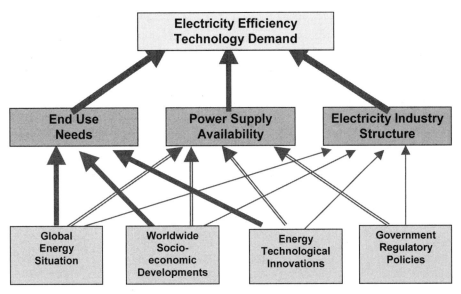

Figure 5.1. An Simple Example of Hierarchical Influence Tracing System

5.3.3 Analytic hierarchy process

Similar to the Hierarchical Influence Tracing System, the hierarchical structure of AHP can also be used for forecasting purposes. For example, it can be applied to estimate the relative probabilities of success (i.e., the probabilities of success in relation to one another with the total summing to 1) or market shares of two competing consumer technologies A and B. In this case, we assume that the probability of success of a consumer technology is determined by four major factors: Affordability, Quality, Timing, and Attractiveness. Then the relative probabilities of success of competing technologies can be developed through the hierarchical structure below:

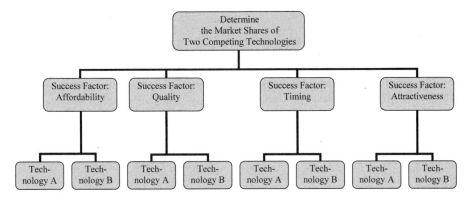

Through the Analytic Hierarchy Process, we will first develop a pair-wise comparison matrix to estimate the relative degree of importance or weight of each

Success Factor in its contribution to the relative probability of success or Market Share of a consumer technology. Then, we will develop a pair-wise comparison matrix of the competing technologies for each Success Factor and use it to estimate the ratings of the technologies with respect to that particular Success Factor. These weights and ratings are then integrated to provide the final relative probability of success or Market Share for each technology.

For example, through matrix analyses, the relative contributions of Affordability, Quality, Timing, and Attractiveness and the ratings of the two technologies with respect to each success factors are given below. Then the probabilities of success of Technologies A and B can be estimated as $0.1(0.3)+0.2(0.4)+0.4(0.7)+0.3(0.2)=0.45$ and $1-0.45=0.55$ respectively.

Success Factor		Affordability	Quality	Timing	Attractiveness	Success Probability
Weight		0.1	0.2	0.4	0.3	
Technology	A	0.3	0.4	0.7	0.2	0.45
Ratings	B	0.7	0.6	0.3	0.8	0.55

5.3.4. Major advantages, pitfalls, and applicability

The major advantages of this assumption are in its rationality and relationship structure. It provides a logical and organized perspective on the reasoning process of the decision maker about these relationships.

The major pitfalls of this assumption are:

1. The relationship structure development is generally very qualitative and often ambiguous.
2. It is difficult to include feedback relationships.
3. For the Hierarchical Influencing Tracing System, changes in the fundamental forces are difficult to determine.

Because of these pitfalls, approaches based on this assumption are generally used for developing long-range insights on technologies and their business environment.

5.4. An Alternative Integrated Approach: Understanding the Future Through Decision-Focused Scenario Analysis

Discussions in Chapter 4 and earlier parts of this chapter have indicated that determining and forecasting the relationships between technology alternatives and decision maker's values are difficult because of a fundamental lack of understanding the underlying mechanism of change in the business, economic, and socio-political environment. Furthermore, although many assumptions can be made about this underlying mechanism, they are generally of questionable accuracy and validity. In response to these difficulties, a more effective approach to manage future uncertainty is the Decision-Focused Scenario Analysis.

5.4.1. Basic characteristics

Scenario analysis is *conceptually different* from traditional forecast or sensitivity analysis methods. Strictly speaking, it does *not* develop a single forecast. Instead, as depicted in Figure 5.2, it develops a set of *structurally different but plausible alternative scenarios* that provides an *envelope to uncertainty* in the future business environment affecting the development or application of technologies of interest. Specifically, in contrast to traditional technology forecasting, decision-focused scenarios are:

- not predictions, but rather are descriptions of alternative plausible futures
- not variations around a mid-point or base-case forecast, but rather are structurally different views of the future
- not generalized views of desired or feared futures, but rather are specifically decision-focused visions of the future
- not products of outside futurists, but rather are the results of management and senior staff's insights and perceptions about the future

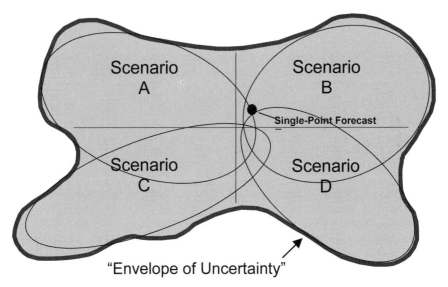

Figure 5.2. Scenarios Form an Envelope to Encompass Major Future Uncertainties

Furthermore, this approach has the following special characteristics:

- It *emulates* effective human reasoning and decision making processes by using a *hierarchical, logical* approach to identify *key factors* in the business environment affecting decisions, and then systematically extract the key uncertainties in these factors as the basis for future scenarios.
- It is *total system-oriented, context-based, inclusive* of all other methods and *integrative* of all available knowledge.

- It is based on *diverse, informed, and reasoned human judgments* (obtained through structured brainstorming and systematic integration of expert knowledge from diverse backgrounds) supplemented by *in-depth research to clarify as much as practical important uncertainties* (in the form of focused research papers).
- It uses *systematic clustering and condensing of factors at each stage* to integrate available knowledge and understanding.
- It uses *repetition and redundancy* to ensure comprehensiveness throughout the scenario development process, to provide constant opportunities to recapture any missing factor or to modify existing factors.

In a complex and dynamic business environment like that of a technology development or application, the decision-focused scenario analysis can be an effective forecasting method because it offers the following advantages:

- Focus on decision objectives
- A total system view
- Rich context on plausible alternative futures
- A basis for managing uncertainty in the business environment

On the other hand, local system-oriented single realization point forecast, even with sensitivity analysis, is generally not only almost impossible to correctly predict the future, but also can be misleading in providing a false sense of accuracy.

5.4.2. Scenario development process

The decision-focused scenario development process consists of the following six iterative steps as summarized in Figure 5.3:

1. Develop the **Decision Focus**, which pinpoints the technology choices that need to be made.
2. Identify **Key Decision Factors**, which are the *key issues in the external environment* that directly affect the decision to be made and need to be forecast for the technology development or application; they often include
   — Technology development
   — Market demand growth
   — Industry structure
   — Government regulations
   — Resource requirements
   — International relations
3. Identify **Major External Drivers**, which are *major forces of changes* in the external environment and the main causes of future uncertainty. In particular, those forces with high importance and low uncertainty will be the *dominant forces* present in all future scenarios.
4. Consolidate the *critical forces,* i.e., those forces with high importance and medium to high uncertainty and those forces with high uncertainty and medium to high importance to form **Axes of Uncertainty**.

5. Use the extremes of these axes to develop *plausible, structurally different, and internally consistent* major **Scenarios of the Future**.
6. Assess **Scenario Implications**, which are *preliminary assessment of the general impacts* of the scenarios on Key Decision Factors and eventually the Decision Focus.

These Scenarios of the external environment may be refined by modifying the key Decision Factors and additional iterations of the scenario development process.

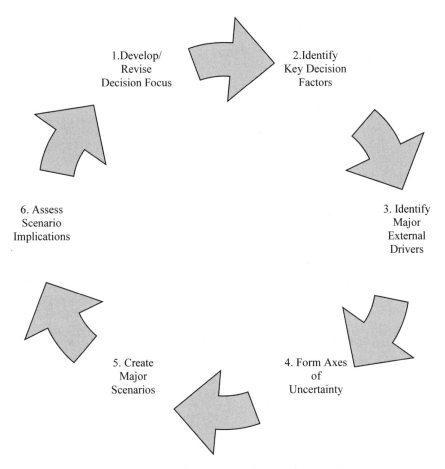

1.Develop/ Revise Decision Focus

2.Identify Key Decision Factors

6. Assess Scenario Implications

3. Identify Major External Drivers

5. Create Major Scenarios

4. Form Axes of Uncertainty

Figure 5.3. The Iterative Process of Scenario Development

5.4.3. Resource requirements

Decision-focused scenario development and analysis is highly *resource intensive*. Formal scenario development for a large business organization or government agency generally requires:

- *Commitment by the chief executive* to a development process of about 3 to 6 months in time length involving 3 or 4 facilitated meetings of major stakeholders in technology development or application
- About 8 to 12 *major stakeholders with diverse background* who will *directly* participate throughout the process *with continuity,* as any interruption will seriously damage consensus building that is essential for scenario development
- An effective *facilitator* to stimulate discussions and provide structure for the scenario development process
- A competent *recorder* to faithfully capture the contents and insights of the scenarios
- Outside *business and technology experts* to supplement the knowledge of the stakeholders
- A professional *research staff* to provide additional detailed studies needed to reduce the uncertainty as much as practical for the major driving forces

Although the resource requirement for a full scenario development is large, it is important to note that the general principles and key elements of the process of scenarios forecasting are applicable to all technology portfolio planning decisions regardless of the level of available resources. Furthermore, scenarios generally tend to be qualitative and descriptive when the decision factors and external drivers are broad. However, scenarios can also be quantitative and specific if the key uncertainty involves two diversely different quantitative models about future changes in the business environment.

5.4.4. A real-world application

Among the numerous applications of the decision-focused scenario forecasting process conducted by the author, a particularly interesting one was the application in the mid-1990s to the decision by the government of an Asian country X for the potential development of virtual reality (VR) technology. Details of this application is presented here as an illustrative example for the scenario forecasting process.

Decision Focus:

This scenario forecasting effort is designed to assist Country X's information technology (IT) industries in making technology investment decisions for the 1996–2003 time period. The focus of the scenario technology forecast is:

"To enter the international virtual reality (VR) related products market and develop add-on value in hardware products for Country X's computer industry by the year 2000. To realize the return on investments by the year 2003."

Scenario Scope:

Development and adoption of virtual reality hardware and software that provides intuitive and natural interactions with computer generated data.

Major Elements of the Decision Focus:

- What opportunities should be pursued?
  - — Application focus
  - — Product focus
- How should the opportunities be accessed?
  - — Market strategy
  - — Technology strategy
  - — Key alliances
  - — Leader organization
  - — Actions and timing needed to realize the desired returns by 2003
- What will be the core technologies?
- What should be the government and private industry funding budgets?
- What should be the return on investment?

Key Decision Factors:

The most important external factors that must be considered in making virtual reality investment decisions include:

- How will the VR technology evolve?
  - — Time frame
  - — Standards
  - — Performance requirements to be met
- What will be the biggest hardware and software market demand?
  - — Standards
  - — Killer application(s)
  - — Benefits to society
  - — Performance requirements to be met
- What will be the industry structure for virtual reality?
  - — Competitors
  - — Leading vendors
  - — Profit locations
- What will be the growth dynamics of Country X's IT industry?
- What will be computing power on the PC platform?

Examples of High Importance and Low Uncertainty External Drivers:

- Computer power on PC (Moore's Law will continue)
- Internet development (will continue to expand)
- Bill of industry upgrades (will happen)
- Computer graphics (will evolve)
- Market information availability (will be available)
- Need for diversions & leisure activities (will grow)
- Investment in 3D graphics by game machine and content providers (will occur)
- Desire for fantasy with real feel and touch (will be important)
- Entertainment applications (will be the main applications)
- Patent trend/barriers (protectionism will increase)
- Country X government sponsored VR related projects (will be made)

- Long term R&D commitment (will be there)
- Country X peripheral manufacturing (will occur)
- Need for stress relief by sensual stimulation (will be important)
- International technology sources (no access to Country X)
- Hazardous situation training applications (will occur)
- U.S. Consumer Products Agency (will take strong measures on health and safety)

Examples of Critical Source of Uncertainty in External Drivers:

- High Importance and Medium Uncertainty External Forces
  — Key components development
  — Killer applications
  — Market size of hardware and software
  — Market size of entertainment applications
  — Niche products
  — Profitability of players
- High Importance and High Uncertainty Forces
  — Interest of venture capitalists
  — Rival countries' technology investment policies
  — Internet development characteristics
  — Value of VR building blocks
  — Worldwide technology diffusion
  — Interrelationships among companies
  — Cost/performance
  — Health and safety factors
- Medium Importance and High Uncertainty Forces
  — Commercial time frame
  — VR research in US/Europe/Japan
  — Country X's internal political situation
  — Japan, Singapore and other Asian country policies and actions
  — VR standards
  — Interest of content providers
  — Country X's system integration capabilities

Note that the above lists of Critical Source of Uncertainty in External Drivers were obtained after *in-depth studies* had been conducted resulting in a set of focused research papers to provide further understanding and to *reduce* as much *uncertainty* as practical for the following important topics:

- Key Components of Virtual Reality
- Needs and Market Potential for Virtual Reality
- Key Player Activities and Status
- Virtual Reality Technology Roadmap and Application Potentials
- Killer Application Characteristics
- Market Value of Virtual Reality Business
- Virtual Reality Standards

Axes of Uncertainty:

For this application, the Critical Sources of Uncertainty in External Drivers were organized into four major *Axes of Uncertainty* for the Virtual Reality business:

1. Industry Structure
   — Investment and source
   — Openness of structure
   — Market barriers
   — Opportunity access
   — Success potential
2. Asia-Pacific Econo-Politics
   — Political conflicts
   — Economic competition
3. Technology Evolution
   — Technology standards
   — Technology integration
   — Technology access
   — Manufacturer dominance
4. Market Demand
   — Cost/performance
   — Leverage
   — Applications
   — Enabling device
   — Safety issues

For each axis of uncertainty, two plausible extremes were developed from specific rationale. The following is an example:

Plausible Extremes of the VR Industry Structure Axis

- *Extreme 1:* Combative and fragmented, due to
  — low investment mainly from corporations
  — protective structure with high market barriers
  — limited access to opportunities
- *Extreme 2:* Cooperative and integrated, due to
  — heavy investments and many from venture capitalists
  — open structure with international cooperation
  — full access to opportunities and many small companies have major successes

The two extremes for each of the other three axes are:

- *Asia-Pacific Econo-Politics:* Open vs. Closed
- *Technology Evolution:* Stuck and/or Disjointed vs. Breakthroughs
- *Market Demand:* Expensive Specialized vs. Cheap Mass Market

These extremes will then result in $2^4=16$ possible combinations of candidate scenarios.

Final Scenario Selection:

Based on the following guidelines:

- Each scenario should be "structurally" different
- Each scenario should be internally consistent to naturally fit into a "story line"
- Each scenario should be plausible
- Each scenario must have decision making utility as a "test bed" for assessing alternative future actions
- Together, the selected scenarios should span the realm of plausible future worlds, or the *"envelope of uncertainty."*

and through extensive discussions among the participants, three final scenarios were selected together with their respective summarizing titles as shown below:

	Axis of Uncertainty			
Final Scenario	Industry Structure	Asia-Pacific Econo-Politics	Technology Evolution	Market Demand
Life in Hell	Combative and Fragmented	Closed	Stuck and/or Disjointed	Expensive and Specialized
Left Behind	Cooperative and Integrated	Closed	Breakthrough	Cheap Mass Market
Waiting for Technology Spring	Cooperative and Integrated	Open	Breakthrough	Cheap Mass Market

For each final scenario, the summarizing title is developed to provide a memory trigger for the participants and a short-handed description of the scenario contents for other interested parties. In addition, a detailed narrative is written for each final scenario to provide a rich and comprehensive description of the unfolding of the scenario.

Example of the Detailed Narrative: the Life in Hell Scenario

Uncertainty Axes: combative/fragmented industry structure, closed Asia, disjointed technology development, expensive specialized applications

Summary: VR is perceived as an emerging technology with important applications. Major countries such as UK, US, and Japan invest vast resources in VR related research in 90's. Although many breakthroughs, the cost performance of VR is still far from the reach of the mass market. Systems can only be afforded in special applications such as flight simulation and industry design. Country X's IT industry becomes interested in VR technology because it provides integration of many technologies and offers good profit potential for a well-structured PC industry.

However, Country X has trouble obtaining key technologies from other countries or developing its own due to patent barriers and the competitive situation. Besides, relations with its political rival country get worse after 1996. Foreign capitalists are reluctant to invest in Country X owing to unstable political future. IT industry suffers because of the lack of investments and struggles to survive.

Detailed Description:

After earthquake disasters in 1996 and 1997 respectively, the Japanese and San Francisco city governments co-sponsor a project to develop an earthquake simulation system using VR technologies. This project is finished by VRtech Inc. in 1998, and includes a system equipped with a 2K x 2K pixels CRT-based HMD, 3D sound, trackers, sensors and a lot of scenarios to train the public in how to survive an earthquake. Users are astonished by the scenes they see and impressed by VR technologies. However, users still complain that the HMD is too heavy, and they feel dizzy while using this system. "These problems are hard to solve," said Tony Simpson, vice president of VRtech, "Because a LCD-based HMD is limited in resolution for its complexity and VR sickness is an unknown territory that still needs to be conquered."

In 2003, Better Corp., a well-known company in Country X, withdraws from the VR industry after reconciling with Sony over the misuse of Sony's VR patents. Better Corp. is interested in VR technologies and has put tremendous resources in research. However, after many years of effort, Better Corp. finds it difficult to buy key technologies of VR or to develop new technologies without violating VR related patents that are held by UK, USA and Japan. Moreover, the relations between Country X and its political rival country are getting worse. As a result, capitalists shift their investments to Malaysia, Vietnam and other countries. Not only Better Corp. suffers these technology problems, but the whole IT industry in Country X also struggles to survive.

In and around the Asian region, the development of VR technology is progressing slowly. They do not completely give up VR and related programs because potential applications are promising. With local government assistance, a few Asian companies obtain funds to develop components for VR, including smart sensors, 6 DOF joysticks, 3D mice, and medium resolution, high-throughput rate graphic cards. Meanwhile, research institutes cooperate with local universities to focus on sensor design, human reaction simulation, and war game simulation using VR technology. With the slow progress of PC-based systems, special applications still use workstations for graphic-intensive tasks. Most observers forecast PCs won't replace workstations for flight-simulation like applications until 2005 or beyond.

By 2000, computing power and 3D graphics have improved significantly. Intel introduces a new graphics accelerator chip for PCs with 5M polygons's speed. It makes VR more realistic and useful in special areas, such as flight simulation, medical surgery practice, etc. However, VR still is not widely used in homes. One important issue is that no standards exist in VR. Since system integration is cumbersome to realize without standards, the prices cannot be cut down. VR sickness is another issue, especially after the University of Edinburgh announces that HMDs cause physiological and psychological impairment if overused, such as 2 hours a day.

Conclusions

Country X used the application results to develop a *robust* VR technology investment strategy that could effectively respond to all three final scenarios. The process was so well received that a formal training program has since been established to the application of the scenario forecasting process to other technology investment decisions.

5.4.5. Insights, caveats, and recent advances

The purpose of scenario forecasting is to develop major plausible futures based on the extremes of unresolvable uncertainties in the external environment. Thus, by definition, all scenarios are about *equally likely*, and there generally is no most likely scenario. Because of this equally likely characteristic, technology portfolio planning needs to be responsive to all these major scenarios.

The experience gained from such applications has also provided the following additional insights for the process:

- Because the scenario development process heavily involves group dynamics, powerful *facilitation skills* are essential for effective implementation of the approach; an experienced, unbiased, and capable facilitator and a faithful recorder are critical to the success of the process.
- The approach is basically a *democratic process to integrate diverse views*, which is the key to a holistic understanding of the future; thus any dominance of views by rank or personality of the process participants must be avoided.
- The uninterrupted participation of the senior executives and professionals throughout the process is essential to ensure the continuity and ownership of the process.
- The scenario development is to integrate all available information and knowledge about future changes in the external environment; thus, it should collect as *much information* and be as *rigorous* as practical.
- To be effective, the scenarios must be eventually *linked to decisions and strategies that provide robust responses to the final scenarios.*

Finally, there have been important recent advances in the implementation of the scenario development process through the use of artificial intelligence-based computer-assisted software, such as the Angler™ developed by SRI International. The general underlying principle here is to provide a computer platform for the interactions among participants, and a data processing and analysis capability for the automated integration of information generated in the process. Specifically, through the computer, the participants can communicate and interact with one another to identify numerous decision factors without time and location restrictions. Each participant can then subjectively classify these factors into various categories. The computer will use data processing and statistical tools, such as pattern recognition and correlation analysis, to combine these subjectively developed categories into a small number of major categories. After the participants review, collectively modify, and reach agreement about the final major categories, they will rate the levels of importance and uncertainty of each category. The computer combines these ratings through pre-agreed algorithms, including arithmetic averaging and geometric averaging, to determine the final level of importance and uncertainty of each category. Such software increases the efficiency of the scenario development process through improved communication and integration of ideas generated in the brainstorming activities. It also keeps a complete record of the process implementation that can be used to provide details of the individual inputs and the development of the axes of uncertainty.

5.4.6. Major advantages, pitfalls, and applicability

The decision-focused scenario analysis provides an effective approach for developing an envelope to encompass major plausible future uncertainties in the business environment for technology development or application. Through repeated brainstorming among diverse experts and in-depth research on technical issues, the approach systematically integrates all available knowledge and understanding about the business environment to provide insights about the future with logically developed diversely different scenarios. These scenarios then form the basis for developing a set of responsive technology portfolio planning strategies.

Nevertheless, a formal application of this approach is resource intensive. Practical experience indicates such an application would require about US$500,000 of professional fees for process facilitation, technical research, and scenario summary over a 6 months period of time. In addition, it will require 40 hours of time from each participating senior executive, and another 80 hours from the staff members of each of these executives. Thus, a formal application is generally appropriate only for large organizations with $1 billion or more in revenue or budget.

However, it should be noted again that the general principle of the approach for integrating diverse understanding about the business environment to develop major future scenarios as planning contingencies is both intuitively sound and applicable to all sizes of organizations and budgets. Furthermore, scenario analysis can be quantitative if the uncertainty lies mainly in values of the input variables or the functional forms of the analytical models.

5.5. References

Delphi Method

Enzer, S., *Delphi and Cross Impact Techniques*, Institute for the Future, 1970

Helmer, O, *The Delphi Method for Systematizing Judgments About the Future*, University of California, Los Angeles, 1966

Linstone, H., *Multiple Perspectives for Decision Making*, Elsevier, 1984

Cross Impact Analysis

Enzer, S., *Cross Impact Techniques in Technology Assessment,* Institute for the Future, 1971

Lipinski, H., *Cross Impact Analysis,* Institute for the Future, 1978

Hierarchical Influence Tracing System

Yu, O., *Scenarios and Strategies: Meeting the Challenges of Global Changes in the Electric Utility Industry, Business Intelligence Program,* D94–1816, SRI International, March 1994

Analytic Hierarchy Process

Saaty, T. and L.Vargas, *Prediction, Projection, and Forecasting*, Springer, 1991.

Scenario Analysis

Ringland, G., *Scenario Planning: Managing for the Future,* Wiley, 1998

Rodriguez, A. and T. Boyce, J. Lowrance, and E. Yeh, E. "Angler: Collaboratively Expanding Your Cognitive Horizon," *International Conference on Intelligence Analysis Proceedings,* MITRE, May 2005

Schwartz, P., *The Art of the Long View: Planning for the Future in an Uncertain World,* Currency, 1996

5.5. Exercises

Problem 5.1

Identify the major leading indicators for forecasting the demand growth of a technology of your interest. Justify your selection of these indicators and develop a forecast for the demand growth of the technology in the next 5 years.

Problem 5.2

Develop a cross-impact matrix for a technology of your interest and use it to develop a 5-year forecast of demand growth in the technology.

Problem 5.3

Develop a hierarchical influence tracing system for a technology of your interest and use it to develop a 5-year forecast of demand growth for the technology.

Problem 5.4

Use the Analytic Hierarchy Process to develop the structure for two competing consumer technologies and estimate their respective market shares in the next 5 years.

Problem 5.5

Develop a set of major axes of uncertainty for a technology of your interest and use it to develop a set of scenarios for the future changes in the technology business environment in the next 5 years.

6 Find the Optimal Technology Portfolio: Major Deterministic Methods

After the decision maker has articulated and assessed the values, identified relevant technology and resource allocation alternatives, and determined the relationships between these alternatives and values in a technology portfolio planning process, he or she will need to find the best resource allocation among the technologies that maximizes the total value of the portfolio. How to efficiently search for the best choice among many alternatives is a challenge to creativity. In many cases, the effective search techniques are closely related to the characteristics of the relationship between alternatives and values, such as those of the following:

1. whether the relationship is deterministic, i.e., without uncertainty, or the effects of uncertainty need to be explicitly considered
2. whether the relationship has a special functional form
3. whether the possible alternatives can be classified by a special scheme

In this chapter, we will focus on a number of major concepts and tools for finding the optimal choice when the relationships between the alternatives and values are deterministic. This deterministic nature can come about either because there is little uncertainty in the relationship, such as in the case where each alternative has a single well-specified impact or value, or because the effects of uncertainty have been absorbed through the use of expected values.

Specifically, Sections 6.1–6.3 will present simple, easy-to-use methods for the cases where the values of alternative technologies have been estimated in an aggregate manner and the number of alternatives to be considered is relatively small and can each be directly evaluated. Sections 6.4–6.7 will present innovative methods for finding the optimal choice for the cases where either the value of the technology portfolio is a concave or linear function of the infinite possible combinations of available alternatives, or the number of possible alternatives are finite but it is computationally prohibitive to directly evaluate all alternatives.

6.1. Aggregate Approach 1: Benefit-Cost Ratio Method

A simple but widely used method for selecting the best technology portfolio is the benefit-cost ratio method that compares the aggregated benefits and costs of various candidate technologies.

6.1.1. Algorithm

The method can be carried out through the following steps:

1. Estimate the total (expected) benefit of each candidate technology and the cost required for developing or applying the technology in the planning period. Reflecting the decision maker's values, the total benefit may be the present worth of the expected net future profits or, more generally, an integrated figure of merit that includes intangible benefits, such as positive contributions to the society and environment. Note that all the effects of future uncertainty on the benefit are absorbed through the use of expected value. Furthermore, the benefits can be estimated through either the decision maker's judgment or sophisticated models.
2. Rank the technology alternatives in decreasing order of the benefit-cost ratio so that Technology i with total benefit B_i and cost required in the planning period C_i, will have the benefit-cost ratio, B_i/C_i, being the ith highest ratio.
3. For a total available budget of C in the planning period, the first n technologies with the sum of C_i equal to or just below C will then be the optimal choice for the portfolio in the planning period.

6.1.2. Illustrative example

As a simplified example, the hypothetical total future benefits and costs in the planning period of 6 electric power technology developments are ranked in decreasing order of benefit-cost ratio as follows:

Technology	B, Total Benefit ($ million)	C, Cost in the Planning Period ($ million)	B/C
Coal	75	15	5.0
Hydro	18	4	4.5
Natural Gas	36	9	4.0
Nuclear	50	20	2.5
Wind	6	5	1.2
Solar	5	5	1.0

If we assume no benefit from fractional investment, then for a $50 million budget in the planning period, the best portfolio will consist of coal, hydro, natural gas, and nuclear for a total cost of $48 million and a total benefit of $179 million, with $2 million remaining unallocated.

6.1.3. Major advantages and disadvantages

This method is appealing in its simplicity and ease to use.

However, it is often controversial in application because the benefit and cost are generally too aggregated and lumpy to reflect the dynamic nature of technology development or application over time. In particular, it does not consider the possibility of partial benefits from partial funding. In the above example, a higher total benefit can be achieved if the remaining $2 million can be invested in wind technology for partial benefit. In reality, ad hoc manipulations will be used to expend the remaining funds.

Furthermore, estimation of the total benefit or the integrated figure of merit can be difficult. As discussed in Chapters 4 and 5, even elaborate economic system and market penetration models often fail to accurately describe the relationships between alternative technologies and their monetary benefits. This difficulty holds especially true for emerging technologies and long-term research where little information about future economic, environmental, and socio-political impacts is available.

The integration of the monetary values with other less tangible values into a single figure of merit can also be complex. The next two methods present systematic but judgmental approaches for developing this integrated figure of merit.

6.2. Aggregate Approach 2: Multiple Rating Evaluation

This approach applies repeatedly the Simultaneous Rating approach in value quantification discussed in Section 2.1 to provide a systematic procedure for estimating the integrated figure of merit as the total benefit for a technology.

6.2.1. Algorithm

The method is accomplished through the following systematic steps:

1. Let R_j be the rating or weight, i.e., degree of relative preference, of Value j among m values the decision maker has determined through the Simultaneous Rating approach in Section 4.1.
2. Use the Simultaneous Rating approach again to obtain r_{ij}, the rating of Technology i among the n technologies, with respect to Value j. If Value j is a monetary unit, and the decision maker is risk neutral, i.e., the value of money is proportional to the amount of money, then r_{ij} can be the monetary benefit of Technology i, as estimated by economic system and market penetration models.
3. Compute $M_i = \Sigma_j r_{ij}R_j$ as the integrated figure of merit or total benefit for Technology i.

If cost is not included in this evaluation, then M_i can be used as the total benefit of Technology i in the application of the Benefit-Cost method.

6.2.2. Illustrative example

Following the example in Section 6.1, let the ratings or weights of two major values, the monetary benefit and environmental impact, be 0.7 and 0.3 respectively. Given the respective ratings of the 5 technologies with respect to these two values, the integrated figures of merit for these technologies are summarized below:

Technology i	Rating r_{ji} with respect to		
	Monetary Benefit (R_1=0.7)	Environmental Benefit (R_2=0.3)	Integrated Figure of Merit $\Sigma_j r_{ji}R_j$
Coal	0.40	0.10	0.31
Natural Gas	0.20	0.20	0.20
Nuclear	0.10	0.15	0.12
Solar	0.05	0.40	0.16
Wind	0.05	0.30	0.13
Hydro	0.10	0.20	0.13

In this case, the benefit-cost ratios will be different from those in the previous example, which may affect the optimal choice for the technology portfolio.

6.2.3. Major advantages and disadvantages

This method is also appealing for its simplicity and ease of use.

However, a major shortcoming is in the fact that simultaneous rating is often imprecise and sometimes inconsistent.

To illustrate this point further, let the three values for evaluating technologies to be Profitability, Quality, and Prestige. Let the simultaneous ordinal ratings of these three values be High (H), Medium (M), and Low (L) respectively. It seems reasonable to assign a numerical rating of 9 to H, 6 to M, and 3 to L. Then the normalized numerical ratings of these values are respectively: $9/(9+6+3)=9/18=0.5$ for Profitability, $6/18=0.333$ for Quality, and $3/18=0.167$ for Presitge.

Suppose there are three candidate technologies with the following characteristics:

- Technology A is highly innovative but untried, so the profitability is likely to be low, but quality is high and the prestige it brings is also high
- Technology B is a middling technology with medium profitability, quality, as well as prestige.
- Technology C is an existing technology for which profitability is high, but quality and prestige are both low.

Then, we have the following ordinal rating matrix:

Technology	Profitability	Quality	Prestige
A	L	H	H
B	M	M	M
C	H	L	L

In turn, we have the following normalized numerical rating matrix and figures of merit:

Technology i	Profitability ($R_1=0.5$)	Quality ($R_2=0.333$)	Prestige ($R_3=0.167$)	Figure of Merit $\Sigma_j r_{ji}Rj$
A	0.167	0.50	0.50	0.333
B	0.333	0.333	0.333	0.333
C	0.50	0.167	0.167	0.333

Now, the figures of merit of three technologies are indistinguishable from one another. It may be possible that the three technologies are indeed indistinguishable, but more likely the identical final figures of merit result from the imprecision of simultaneous ratings. In addition, there is no formal procedure to check the internal consistency of these ratings.

6.3. Aggregate Approach 3: Analytic Hierarchy Process

As presented in Chapter 2, Analytic Hierarchy Process (AHP) uses the hierarchical structure, the observation that humans can more precisely differentiate two items than three or more items at one time, and the matrix theory to develop a systematic and consistent process for comparing and assessing values. The application here is to use AHP repeatedly not only for value assessment but also for alternative evaluation.

6.3.1. Algorithm

The algorithm for AHP has been outlined in Chapter 2. However, in making the optimal choice, we will need to expand the hierarchical structure. The simplest example will have the hierarchical form shown in Figure 6.1.

- **Hierarchy 1:** Determine the Integrated Figures of Merit for Candidate Technologies
- **Hierarchy 2:** Assess values through pair-wise comparison of the degree of preference by the decision maker
- **Hierarchy 3:** Evaluate candidate technologies with respect to each value through pair-wise comparison of the degree of relative preference by the decision maker

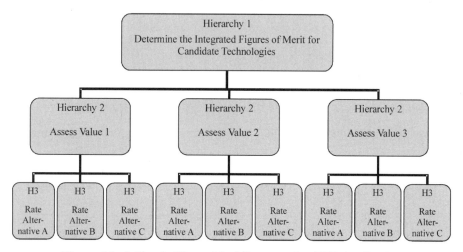

Figure 6.1. A Simple Hierarchical Structure for AHP Application

Additional sub-hierarchies can be added to Hierarchy 2, if the values are further subdivided into component values of successive levels of detail.

6.3.2. Illustrative example

We will continue the example above, which was also discussed in Section 2.2, in which the three values for evaluating technologies, Profitability, Quality, and Prestige, have already been rated through AHP as 0.71, 0.22, and 0.02 respectively.

Now we assume that the pair-wise comparison matrices of the three candidate technologies with respect to the three values have been developed by the decision maker respectively as follows:

	Profitability			Quality			Prestige		
	A	B	C	A	B	C	A	B	C
A	1	1/5	1/9	1	3	6	1	4	9
B	5	1	1/3	1/3	1	3	1/4	1	3
C	9	3	1	1/6	1/3	1	1/9	1/3	1

Following AHP procedures, the ratings r_{ij} of technology i with respect to value j, together with the integrated figures of merit as well as the consistency indices of the pair-wise comparison matrices are summarized below:

	Profitability	Quality	Prestige	Integrated Figure of Merit
Weight (R_j)	0.71	0.22	0.07	$\Sigma_j r_{ij} R_j$
Rating (r_{ij})				
A	0.06	0.65	0.73	0.24
B	0.27	0.25	0.20	0.26
C	0.67	0.10	0.07	0.50
CI	0.015	0.01	0.005	

Different from the results of the Multiple Rating Evaluation approach, Technology C, which is the best choice, is now well differentiated, rather than indistinguishable, from the other two technologies.

6.3.3. Major advantages and disadvantages

AHP is attractive and widely applied because of its hierarchical structure, simplicity, precision, and consistency. Furthermore, empirical evidence has shown that the pair-wise preference comparison matrix is highly robust and stable in that small inconsistency generally does not significantly affect the weights or ratings of the preference obtained through the row averaging process.

However, the integrated figures of merit are often still too aggregated to reflect the dynamic nature of technology development or application. The top-down hierarchy is also difficult to incorporate feedbacks from lower echelons of the hierarchy.

As mentioned in Chapter 2, another potential pitfall is that the 1–9 scale is not a linear representation of the degree of preference, which may cause distortions in the numerical representations of the pair-wise comparisons and the integrated figures of merit.

6.4. Increasing and Concave Value Function: Generalized LaGrange Multiplier Method

If the values of candidate technologies are known continuous functions of the amount of resources allocated and each function becomes concave and increases beyond a threshold amount of resources, then a simple and efficient method based on classical calculus can be used to find the optimal allocation of resources among these technologies.

6.4.1. Mathematical foundation

In this method, we assume that the value of each Technology i, $V_i(x)$, $i = 1,2,\ldots,n$, in n candidate technologies is a continuous function of the amount of resource x allocated to the development or application of Technology i. We also assume that V_i's are independent and additive. In other words, the value from investing in one technology is not affected by the investments in other technologies, and the total value is the sum of the individual values. We further assume that V_i's are all *S-shaped*. By this assumption, $V_i(x)$ *increases* with x, but *approaches a plateau* as x approaches infinity. Moreover, V_i becomes *concave* when x exceeds a threshold amount x_{io}. Specifically, if allocating two amounts of resource y and z $> x_{io}$ to Technology i yields respective values $V_i(y)$ and $V_i(z)$, and if $x = py+(1-p)z$ with p between 0 and 1, then $V_i(x) \geq pV_i(y) + (1-p)V_i(z)$.

The problem of finding the best allocation of a total amount of resource B among n technologies can now be expressed mathematically as:

$$\text{Maximize } \Sigma_i V_i(x_i)$$
$$\text{subject to } \Sigma_i x_i \leq B$$
$$x_i \geq x_{io} \text{ for } i = 1,2,\ldots,n$$

Based on the concept of the LaGrange multiplier, this problem is equivalent to:

$$\text{Maximize } \Sigma_i[V_i(x_i)-\lambda x_i]$$
$$\text{subject to } \Sigma_i x_i \leq B$$
$$x_i \geq x_{io} \text{ for } i = 1,2,\ldots,n$$

where λ is the generalized LaGrange multiplier for the resource.

Because V_i's are S-shaped and concave when $x_i > x_{io}$, for a give λ, the set of x_i* $> x_{io}$ maximizes $\Sigma_i[V_i(x_i)-\lambda x_i]$ if x_i* is the value at which the derivative of V_i with respect to x_i, dV_i/dx_i, equals λ. If $\Sigma_i x_i* = B$, then x_i*'s form the best allocation of the total resource B to the n candidate technologies.

6.4.2. Algorithm

Based on the above discussion, the optimal resource allocation can be found through the following steps:

1. Start by setting $\lambda = 0$, then x_i* at which $dV_i/dx_i = \lambda$ would be infinity for all i.
2. Gradually increase λ from 0 and determine the correspondingly decreasing x_i*, at which $dV_i/dx_i = \lambda$.
3. As λ gradually increases from 0, $\Sigma_i x_i*$ becomes finite and gradually decreases.
4. Because V_i is continuous, if the total amount resources available, B, is sufficiently large, $\Sigma_i x_i*$ will eventually become equal to B, with each $x_i* > x_{io}$.

5. x_i^*, i=1,2,…,n are then the optimal allocation of $\Sigma_i\ x_i^* = B$ total amount of resource to the n Technologies, i.e., the allocation that maximizes the combined value among the n Technologies.

6.4.3. Illustrative Example

Three candidate technologies, A, B, and C, are under development. Let the value functions of these technologies, which are expressed in millions of dollars of return for the decision maker, to be given respectively as follows:

$VA(x_A) = 40\{1-\exp[-0.6(x_A-3)]\}$ for $x_A>3$, and 0 otherwise
$VB(x_B) = 30\{1-\exp[-0.5(x_B-2)]\}$ for $x_B>2$, and 0 otherwise
$VC(x_C) = 20\{1-\exp[-0.4(x_C-1)]\}$ for $x_C>1$, and 0 otherwise

where exp is an exponential function, x_A, x_B, and x_C are resources in units of thousands to be invested in the three technologies, respectively. These value functions become concave and increasing when the amounts of resources x_A, x_B, and x_C allocated to the three technologies exceed their respective thresholds of 3, 2, and 1 thousands of units.

The three value functions, together with a typical λ and the corresponding values of x_A, x_B, and x_C at which the derivatives of these value functions equal λ, are shown in Figure 6.2.

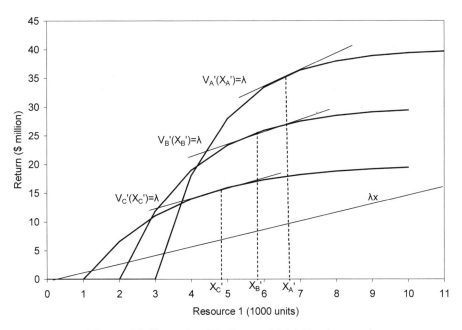

Figure 6.2. Example of LaGrange Multiplier Approach

For a generic value function $V=Vo\{1-\exp[-b(x-xo)]\}$, $x^*=xo-\{1/b\}\log(\lambda/(bVo)$ is the value at which $dV/dx=\lambda$. Thus, applying the Generalized LaGrange Multiplier method, we will gradually increase λ from 0 and solve for the amount of resource allocation to Technology i, x_i^*, until $\Sigma_i x_i^* = B$.

The values of resource allocations, x_A^*, x_B^*, and x_C^* and their sum as functions of λ, are summarized graphically in Figure 6.3.

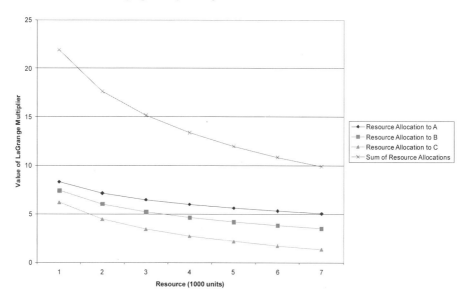

Figure 6.3. Optimal Resource Allocations as Functions of the LaGrange Multiplier

Specifically, for B= \$10 thousand units, we find that the optimal allocation to be

x_A = 5.10 thousand units
x_B = 3.55 thousand units
x_C = 1.35 thousand units

6.4.4. Major advantages and disadvantages

With well-defined S-shaped value functions, this method is extremely efficient and easy to use in finding the optimal allocation of large amounts of resources.

However, the method fails in regions where the value functions that are not concave. In that case, the Dynamic Programming approach of Section 6.7 can be applied.

6.5. Linear Value Function with a Convex Set of Feasible Alternatives: Linear Programming

A special case of the concave value function is the linear value function, where the value is proportional to the resources allocated. Furthermore, if the set of acceptable or feasible alternatives is a convex set, i.e., any linear combination of two feasible alternatives is also feasible, then the search for the optimal alternative becomes a Linear Programming problem, for which a well-established algorithm exists. In fact, because of its wide applications, from resource allocation to network optimization and many others, this special search algorithm, the Simplex Method, as created by George Dantzig, is one of the best known and most used techniques for finding the best alternative in an optimization problem. In this section, we will not present the theoretical details of this algorithm, which are readily available from many sources including those given in the reference section of this chapter. Instead, we will highlight a few major characteristics of linear programming that are particularly relevant to technology portfolio planning.

6.5.1. A simple linear program for a technology portfolio decision

The following is a simple yet typical example of formulating the technology portfolio decision as a linear program:

A decision maker needs to allocate $100 million, 100,000 person-hours of staff time, and 40,000 of management time to two technology projects:

1. improving an existing technology that has an annual return rate of 10% and requires 500 person-hours of staff time but no management time per million dollar of investment
2. developing a new technology that has an annual return rate of 30%, but requires 2000 person-hours of staff time and 1000 person-hours of management time per million dollar investment

Let X_1 and X_2 be, respectively, the amounts in million of dollars of investment for the projects of improving the existing and developing the new technologies. Then the technology portfolio decision can be represented by the following linear program:

$$\text{Maximize the total annual return } V = 0.1X_1 + 0.3X_2$$

$$
\begin{aligned}
\text{Subject to} \quad X_1 + X_2 &\leq 100 \text{ (funds)} \\
0.5X_1 + 2X_2 &\leq 100 \text{ (staff time)} \\
X_2 &\leq 40 \text{ (management time)} \\
X_1 &\geq 0 \\
X_2 &\geq 0
\end{aligned}
$$

6.5.2. A general linear program and its basic characteristics

A general linear program has the following mathematical form:

Maximize $V=C^TX=\Sigma C_j X_j$

Subject to $AX \leq B$ or $\Sigma a_{ij}X_j \leq B_i$ for i=1,2,...,m

                  $X \geq 0$ or $X_j \geq 0$ for j=1,2,...,n

where $X = (X_j)$ is the nx1 vector of the levels of resource-consuming activities

$C = (C_j)$ is the nx1 vector of unit profits associated with the activities

j=1 through n, with C^T being the 1xn transpose vector of C

$A = [a_{ij}]$ is a mxn matrix of a_{ij}, the amount of Resource i consumed by a unit of Activity j

$B = (B_i)$ is the mx1 vector of Resources i=1,2,...,m, available

It is useful to note that by reversing the signs of C and B, this maximization linear program can be transformed to a minimization linear program.

This generic linear program has the following basic characteristics:

1. Certainty: all functions are deterministic.
2. Linearity: all functions for both the value and the constraints are linear in that they are proportional and additive; this also means that the functions are continuous.
3. Non-negativity: all X_j's are non-negative.

6.5.3. Graphical representation of a two-activity linear program

The graphical representation of the linear value function and constraints of a linear program with two activities such as the one in the above example is given in Figure 6.4.

In this graphical representation, we observe the following additional characteristics of a linear program:

1. The set of feasible alternatives, or feasible region as outlined by OABCD, forms a polygon, or polytope in general
2. Because the value function is linear, and the polygon is convex, the maximum of the value function occurs at one of the corner points of the feasible region, where each corner point is the intersection of m of the m+n linear boundary equations of the constraints.

Characteristic (2) indicates that a linear program has a finite number of corner points as candidates for the optimal solution. Since each corner point is the intersection of m of the m+n linear boundary equations, the possible number of corner points is (m+n)!/(m!n!). As a result, for any realistic problem of m and n in hundreds or even thousands, it will be computationally prohibitive for finding the optimal solution by exhaustive enumeration of all possible corner points. Thus, an innovative method to search for the optimal corner point is needed.

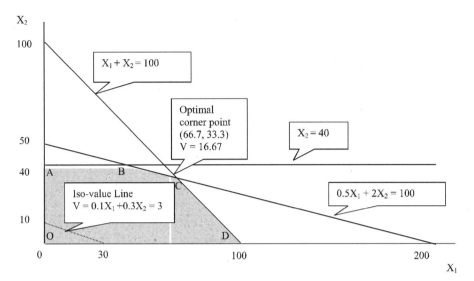

Figure 6.4. Graphical Representation of a Two-Activity Linear Program

6.5.4. Graphical solution to a two-activity linear program

To graphically solve the linear programming problem of finding the optimal alternative that maximizes the value function, we need to construct a typical iso-value line, as represented by the line segment indicated in Figure 6.4 for V=3. In this case, all combinations of (X_1, X_2) on the iso-value line will yield the same value, an annual return of $3 million and the intercepts on the X_1 and X_2 axes are (10,0) and (0,30) respectively. We further observe that the iso-value line for V=6 is parallel to that for V=3 with the intercepts on the X_1 and X_2 axes being (20,0) and (0,60) respectively. Thus, if we continue to move the iso-value line in parallel towards the northeastern direction, the maximum of the value function will be reached at the last corner point before the line leaves the feasible region. Specifically, the maximum of the value function is at $(X_1=66.7, X_2=33.3)$, which is the intersection of the linear boundary equations:

$$X_1 + X_2 = 100$$
$$0.5 X_1 + 2X_2 = 100$$

with V=(0.1)(200/3)+(0.3)(100/3)=\$16.67 million.

6.5.5. Outline of a systematic solution procedure and the Simplex method

The graphical solution procedure works only for a two-activity linear program, as we are not able to visualize the iso-value lines and the convex feasible region of a linear program with more than two activities. However, the underlying concept of the following solution procedure, explained through the graphical representation of the two-activity linear program, applies to linear programs with more than two activities.

In this systematic solution procedure, we may view the values as physical heights. Thus, we may regard the search for the maximum of the value function as finding the peak of a mountain formed by a plane resting on a polygon-shaped base with cliffs along the boundaries of the polygon. With this visual image, a systematic procedure for finding the maximum value, or the peak of the mountain, may be developed as follows:

Step 1.   Start with the origin corner point, (0,0), or any other initial corner point.

Step 2.   Examine whether there is a direction at which the height will increase from this corner point; if none, then we have reached the peak or the corner point with the maximum value.

Step 3.   If there are directions of increasing heights or values, we will proceed in the direction of the greatest increase, which will always be along one of the boundaries of the feasible region, until we reach the end of this boundary line which is a new corner point, and we then go to Step 2.

For the example shown in Figure 6.4, starting with Step 1 at the origin corner point, the direction of the greatest increase for Step 2 is the direction of X_2, which increases the value function by \$0.3 million for each unit increase in that direction, while the direction of X_1 will only increase the value function by \$0.1 million. Following Step 3, we will continue in this direction of the greatest increase until we reach the end of this direction, i.e., another corner point, then we return to Step 2.

This observation of finding the maximum by moving through successively improved corner points is the foundation of the Simplex method developed by George Dantzig for solving linear programming problems. In this method, the corner point, or the solution of m out m+n linear boundary equations of the constraints, continues to evolve through systematically replacing the activity variable currently in the equation ("a basic variable") with the smallest profit coefficient (Cj) by a new activity variable with the largest positive-valued profit coefficient among the activity variables currently not in the equation ("non-basic variables"). The maximum is reached when all variables not in the equation have zero or negative-valued profit coefficients.

6.5.6. Shadow prices of resources and relation to LaGrange multipliers

The shadow price of a resource in a linear program is the incremental change of the value due to one unit change of the resource. In this case,

1.   If we increase the funds available by \$1 million, and re-solve the linear program, we will find the new solution as (X_1=68, X_2=33) with V=\$16.7 million. Thus the shadow price for the funds is the incremental increase in value, \$16.7–\$16.67= \$0.03 million, or 3% of the investment.

2.   If we increase the staff time by 1000 person-hours, and re-solve the linear program, we will find the new solution as (X_1=67, X_2=34) with V=\$16.9 million. Thus, the shadow price of the staff time is the incremental increase in value, \$16.9–\$16.67=\$0.23 million, or \$230 per person-hour.

3. If we increase the management time by 1000 person-hours, and re-solve the linear program, we will find the new solution as ($X_1=66.7$, $X_2=33.3$) with V=$16.67 million. Thus, the shadow price of the management time is the incremental increase in value, $16.67–$16.67=0.

From the above, we observe that if there is surplus for a resource at the optimum, such as the management time, then the shadow price or value added of the resource is 0. In other words, a resource is only valuable if it contributes to an increase in the value function. This observation is important when the technology portfolio decision maker is negotiating with the resource suppliers about their value added to the organization in determining whether additional resources should be employed.

We further observe that the shadow price, which is the slope (incremental change) of the value function at a corner point (resource constraint), is conceptually identical to the generalized LaGrange multiplier discussed in Section 6.5. Therefore, we can also find the shadow prices y_i's of the resources through the following dual minimization linear program of the original maximization linear program:

$$\text{Minimize} \quad U = Y^T B = \Sigma B_i Y_i$$
$$\text{Subject to} \quad Y^T A \geq C \text{ or } \Sigma a_{ij} Y_i \geq C_j \quad \text{for } j=1,2,\ldots,n$$
$$\phantom{\text{Subject to}} \quad Y \geq 0 \text{ or } Y_i \geq 0 \qquad\qquad \text{for } i=1,2,\ldots,m$$

where $Y = (Y_i)$ is the nx1 vector of the shadow prices, with Y^T being the 1xn transpose vector of Y

$B = (B_i)$ is the mx1 vector of resources i=1,2,…,m, available

$A = [a_{ij}]$ is a mxn matrix of a_{ij}, the amount of Resource i consumed by a unit of Activity j

$C = (C_j)$ is the nx1 vector of unit profits associated with the activities j=1 through n, with C^T being the 1xn transpose vector of C

6.5.7. Major advantages and disadvantages

The major advantages of the linear programming problem formation and solution are in its wide applications and well-established solution procedures and commercial software packages. On the other hand, the major disadvantages of linear programming are in its lack of consideration about uncertainty and the requirement for linearity.

These disadvantages can be especially problematical to technology investments, because large uncertainty is inherent in technology development and application. Furthermore, most of the relationships between alternatives and values in the technology portfolio planning process are non-linear. For example, as discussed in Chapter 4, technology demand growth as a function of time and financial return as a function of investment are generally S-shaped. Moreover, as will be discussed in Chapter 7, the combined risk of a technology investment portfolio is, in general, not linearly related to the risks of individual investments. As a result, applications of linear programming to technology portfolio planning and management have been somewhat limited.

6.6. Efficient Search for a Large Number of Sectionable Alternatives 1: Branch-and-Bound Approach

For a linear programming problem, if we add one seemingly innocuous requirement that all feasible solutions be integer-valued, the complexion of the resulting problem, which is generally called Integer Linear Programming, is radically changed because the feasible region is no longer convex, as the linear combination of two feasible, i.e., integer-valued, solutions may not be integer-valued or feasible. Consequently, the Simplex solution method for the continuous-valued linear programming problem can no longer apply. If we simply round off the optimal solution from the continuous-valued linear programming problem, it may be far different from the optimal solution of the integer-valued linear programming problem. Furthermore, even though the number of integer-valued feasible solutions is finite, for any realistic application with hundreds or more constraints, the astronomical number of feasible solutions will again prohibit exhaustive evaluation even by the most advanced computer. Thus, an innovative method for searching for the optimal solution is once again needed. One such method is the branch-and-bound approach.

Applications of the branch-and-bound approach are not limited to the solution of integer linear programming related problems. It is applicable whenever the optimal solution of a special problem of interest is difficult to obtain directly, but the optimal solution of a general problem whose feasible solutions include those of the special problem can be obtained efficiently.

6.6.1. Basic algorithm

The basic algorithm of the branch-and-bound approach for a *maximization* problem is outlined as follows:

Step 1. Let the special optimization problem of interest be S and define a general problem G for which the feasible alternatives of S are a subset of the feasible alternatives of G. As a result, the optimal solution for G may not be S-feasible, i.e., a feasible alternative for S; however, if it is S-feasible, then it is optimal for S

Step 2. Find the optimal solution of G which is set as the upper bound UB of the optimal solution of S. If UB is S-feasible, then it is also the optimal solution of S. Otherwise, set minus-infinity as lower bound LB and go to Step 3.

Step 3. Divide the problem with UB (initially G) into two or more sub-problems or branches B_i's, such that the sets of feasible alternatives for the B_i's are mutually exclusive but they collectively form the set of feasible alternatives of the problem with UB.

Step 4. Find the optimal solution of each B_i. Let the new UB be the largest of *all* the optimal solutions of B_i *including* those of any previous sub-problems that have not yet been divided. Let the new LB be the largest of *all* S-*feasible* optimal solutions of B_i *including* those of any previous sub-problems that have not yet been divided. If no optimal solution of B_i's is S-feasible, then LB remains the same as the previous LB. Now, if UB=LB, then the optimal solution with UB is the optimal solution of S. Otherwise, go back to Step 3.

For minimization problems, we change the largest to the smallest, reverse the roles of UB and LB, and set the initial UB to be plus-infinity if the initial LB is not S-feasible.

6.6.2. Examples of application

Example 1. Integer Linear Programming

For the technology portfolio linear programming problem discussed in Section 6.5, let us add the requirement that each investment must be in whole number of millions and all person-hours must be in whole hours. In other words, it becomes an integer linear programming problem S as follows:

Maximize the total annual return $\quad V=0.1X_1 + 0.2X2$
Subject to
$$X_1 + X_2 \le 100 \text{ (funds)}$$
$$0.5X_1 + 2X_2 \le 100 \text{ (staff time)}$$
$$X_2 \le 40 \text{ (management time)}$$
$$X_1 \ge 0$$
$$X_2 \ge 0$$
$$X_1 \text{ and } X_2 \text{ are integers}$$

Applying the branch-and-bound approach for finding the optimal solution, we start with the continuous-valued linear programming problem discussed in Section 6.5 as the general problem G. The optimal solution $(X_1=66.7, X_2=33.3)$ with $V=16.67$ not integer-valued thus not S-feasible. We set the upper bound UB=16.67, the lower bound LB=minus-infinity, and divide G into two sub-problems or branches with mutually exclusive but collectively exhaustive sets of feasible alternatives (here both X_1 and X_2 are not integer-valued; for simplicity, we choose to divide G along X_1):

(B1): G with the additional requirement that $X_1 \le 66$
Optimal solution is $(X_1=66, X_2=33.5)$ with $V=16.65$ and not S-feasible
(B2): G with the additional requirement that $X_1 \ge 67$
Optimal solution is $(X_1=67, X_2=33)$ with $V=16.6$ and S-feasible

The new UB=maximum of all branches=16.65
The new LB=maximum of all feasible branches=16.6

Because UB does not equal LB, we further divide the branch with UB, i.e., B1 into two sub-problems with mutually exclusive but collectively exhaustive sets of feasible alternatives (in this case, we divide B1 along X_2 because it is not integer-valued):

(B11): B1 with the additional requirement that $X_2 \leq 33$
Optimal solution is (X_1=66, X_2=33) with V=16.5 and S-feasible
(B12): B1 with the additional requirement that $X_2 \geq 34$
Optimal solution is (X_1=64, X_2=34) with V=16.6 and S-feasible

The new UB=maximum of all branches=16.6
The new LB=maximum of all feasible branches=16.6

Since UB=LB, the optimal solution for S is (X_1=64, X_2=34) with V=16.6.
The overall branch-and-bounding process is graphically summarized below.

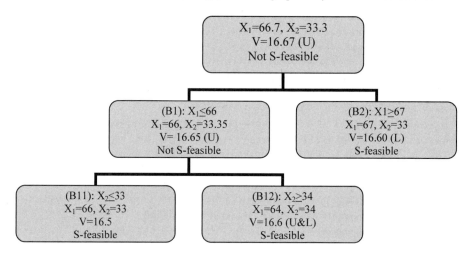

Example 2. Assignment problem[10]

The special problem S is that a decision maker needs to assign 4 staff members
A,B,C,D to be managers for 4 technology projects 1,2,3,4, with the requirement of
one staff member for each project and one project per staff member. The values in
millions of dollars for assigning various staff members to different project are
summarized in the matrix below. The decision maker desires to maximize the value
of the assignment.

Values of assigning staff members to technology projects

---

[10] Assignment problem generally can be solved as a linear programming problem; it is
discussed as a demonstration for the Branch-and-Bound approach.

Project	1	2	3	4
Staff				
A	85	70	60	10
B	6	15	90	76
C	50	80	5	75
D	75	84	82	25

The general problem G is to assign each staff member to the project with the highest value. In this case, the best assignments for G are A1 + B3 + C2 + D2 with value = 85 + 90 + 80 + 84 = 339, but it is not S-feasible because two staff members, C and D, are assigned to project 2. Set UB=339 and LB=minus-infinity.

Divide G into 4 mutually exclusive and collectively exhaustive sub-problems or branches (we can divide G by the assignments of a specific staff member or by the assignments to a specific project; for simplicity, we choose to divide G by the assignments of staff member A):

(B1): Assign A to1 and solve the general problem for the remaining assignments. The best assignments are A1+B3+C2+D2 with value = 85+90+80+84 = 339, but not S-feasible.

(B2): Assign A to 2 and solve the general problem for the remaining assignments. The best assignments are A2+B3+C4+D3 with value = 70+90+75+ 82 = 317, but not S-feasible.

(B3): Assign A to 3 and solve the general problem for the remaining assignments. The best assignments are A3+B4+C2+D2 with value = 60+76+80+84 = 300, but not S-feasible.

(B4): Assign A to 4 and solve the general problem for the remaining assignments. The best assignments are A4+B3+C2+D2 with value = 10+90+80+84 = 264, but not S-feasible.

New UB = Maximum (All branches) = 339
New LB = Max (All feasible branches) = minus-infinity

Since UB does not equal LB, we further divide the branch with UB, i.e., (B1), into three mutually exclusive and collectively exhaustive sub-problems (since now A has been assigned to 1, we will divide (B1) into these sub-problems by the three possible assignments of B to the three remaining projects):

(B11): A1 and B2 and solve the general problem for the remaining assignments. The best assignments are A1+B2+C4+D3, with value = 85+15+75+82=257 which is S-feasible

(B12): A1 and B3 and solve the general problem for the remaining assignments. The best assignments are A1+B3+C2+D2, with value = 85+90+80+84=339 which is not S-feasible

(B13): A1 and B4 and solve the general problem for the remaining assignments.

The best assignments are A1+B4+C2+D2, with value = 85+76+80+84=325 which is not S-feasible

New UB = Maximum (All branches) = 339
New LB = Max (All feasible branches) = 257

Again since UB and LB are not equal, we divide the branch with UB, i.e., (B12), into two mutually exclusive and collectively exhaustive sub-problems:

(B121): A1, B3, C2, and D4 with value 85+90+80+25=290 which is S-feasible
(B122): A1, B3, C4, and D2 with value 85+90+75+84=334 which is S-feasible

New UB = Maximum (All branches) = 334
New LB = Maximum (All feasible branches) = 334

Since now UB=LB, the optimal solution for S is A1,B3,C4, and D2 with optimal value = 334.

The overall branch-and-bounding process is again graphically summarized below.

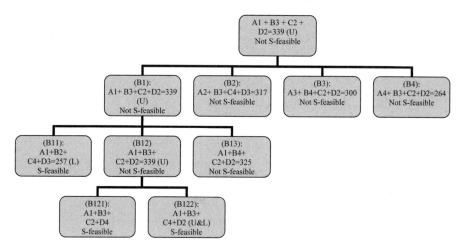

6.6.3. Major advantages and applicability

The major advantage of the branch-and-bound approach is that it greatly reduces the number of computational steps and makes the search for optimal solutions efficient and practical. For example, for the assignment problem of n staff members and n technology projects, if we find the optimal solution through direct enumeration possible assignment, whether they are S-feasible or not, the number of computations will be proportional to n!, as we can sequentially assign staff members to different jobs. This number grows exponentially with n and soon becomes computationally prohibitive. However, in the branch-and-bound approach, the number of computations is greatly reduced because it searches only the most promising branches. In Example 2 above, the direct enumeration method would have taken

4!=24 steps to find the optimal solution, but the branch-and-bound approach takes only 9 steps.

Another major advantage is that whenever we stop after an iteration, we know exactly the upper and lower bounds for the optimal value. For Example 1, if we stop after the first iteration, we know that the optimal value is between 16.6 and 16.65. Similarly for Example 2, if we stop after the first iteration, we know that the optimal value is between 257 and 339. Therefore, for large problems, we can stop after an iteration when the difference between the upper bound and lower bound values reaches a tolerable limit.

However, the effective application of the branch-and-bound approach generally requires the optimization problem to have the following two characteristics:

1. A general problem G exists for which the set of feasible alternatives of G includes those of the special problem S and G can be solved efficiently.
2. The feasible alternatives of G can be easily sectioned into mutually exclusive and collectively exhaustive subsets.

6.7. Efficient Search for a Large Number of Sectionable Alternatives 2: Dynamic Programming Approach

In real-world resource allocation problems for technology portfolio planning, resources are finite and are generally measured in discrete units, such as dollars and person-hours. Theoretically, we can obtain the optimal allocation of resource to technology development or application by systematically computing the overall value of every possible combination of resource allocation and finding the one combination that produces the greatest overall value. However, applying this exhaustive enumeration method for investing m resources, each with M_i units available, into n technologies, will result in the amount of computations proportional to $\Pi_i(M_i +1)^n$. Unfortunately, this amount inceases exponentially with m, M_i, and n, and will quickly become prohibitive in terms of computational time for any practical size technology portfolio planning decisions.

The Dynamic Programming approach, which was developed by Richard Bellman in 1950s, dramatically reduces the amount of computations for allocating limited resources to a large number of technologies by making it increase as an algebraic function of m, M_i, and n. The principle of the approach can be stated almost tautologically as follows:

6.7.1. Dynamic programming principle: the principle of optimality

If $x_A + x_B + x_c = x$ are the optimal allocation of the combined resource x to technologies A, B, and C, with $x_A + x_B = y$ and $x_c = x - y$, then x_A and x_B must also be the optimal allocation of resource y to technologies A and B.

Based on this simple principle, we can develop an efficient algorithm to systematically find the optimal allocation of resources to various technologies. For simplicity of discussion, we will develop this algorithm for only one type of resource. However, it can be easily generalized to include multiple types of resources.

6.7.2. Algorithm for allocating M units of resource to n technologies[11]

M units of resource are available to be allocated to n technologies. Let $V_i(x)$ be the value of allocating x units of resource to Technology i, for x=0,1,.., M, and i=1,2,..,n.

Initiation
Start with Technology 1, for x=0,1,..,M, let

$V_1^*(x)$ = the optimal value from allocating x units of resource to Technology 1, which is identical to $V_1(x)$ when Technology 1 is the only technology under consideration

$x_1^*(x)$ = the optimal amount of resource to be allocated to Technology 1 when x amount of resource is available to Technology 1, which is identical to x when Technology 1 is the only technology under consideration

Recursion
For i = 2,3,…,n–1, and x=0,1,2,…,M, let

$V_i^*(x)$ = optimal value from allocating x units of resource to Technologies 1, 2, …,i

    = Max $_{y=0,1,2,...,x}$ $[V_{i-1}^*(y)+V_i(x-y)]$

$y_i^*(x)$ = the y that maximizes $V_i^*(x)$, which is also the optimal amount of resource to be allocated to Technologies 1,2, …, i–1 together as a subgroup when the total amount of resource available to Technologies 1,2,..,i is x

$x_i^*(x)$ = $x-y_i^*(x)$ = the optimal amount of resource to be allocated to Technology i, when the total amount of resource available for Technology 1,2,…,i is x

End

$V_n^*(M)$ = the optimal value from allocating M units of resources to n technologies

    = Max $_{y=0,1,2,...,M}$ $[V_{n-1}^*(y)+V_n(M-y)]$

$y_n^*(M)$ = the y that maximizes $V_n^*(M)$, which is also the optimal amount of resource to be allocated to Technologies 1,2,..,n–1 together as a subgroup, when the total amount of resource available to Technologies 1,2,..,n is M

$x_n^*(M)$ = $M-y_n^*(M)$ = the optimal amount of resource to be allocated to Technology n, when the total amount of resource available for Technology 1,2,…,n is M

Amount of Computation
(M+1)n additions and comparisons versus $(M+1)^n$ additions and comparisons by the exhaustive numeration method.

---

[11] This algorithm with simplified symbols is based on original work by the author.

6.7.3. Illustrative example

A technology company wants to find the optimal allocation of $10 million in units of $1 million to 3 technologies. Specifically, we have n=3, M=10. Following the dynamic programming algorithm, the value functions of these technologies together with the optimal value functions and optimal allocations for various amounts of resources, all in million dollar units, are summarized in the following series of tables.

Table 6.1. Optimal Allocation for Technologies 1 and 2

Amount Available (in 10^6)	Value Function of Technology 1 (in 10^6)	Value Function of Technology 2 (in 10^6)	Optimal Value Function for Technologies 1&2 (in 10^6)	Optimal Allocation to Technologies 1 when x amount is available to Technologies 1&2 (in 10^6)	Optimal Allocation to Technology 2 when x amount is available to Technologies 1&2 (in 10^6)
x	$V_1(x)= V_1{}^*(x)$	$V_2(x)$	$V_2{}^*(x)$	$y_2{}^*(x)$	$x_2{}^*(x)=x-y_2{}^*(x)$
0	0	0	0	0	0
1	0	0	0	0	1
2	39	0	39	2	0
3	63	44	63	3	0
4	78	79	79	0	4
5	86	106	106	0	5
6	92	126	126	0	6
7	95	143	145	2	5
8	97	155	169	3	5
9	**98**	**165**	**189**	**3**	**6**
10	99	173	206	3	7

Here, $V_2{}^*(x) = \text{Max}_{y=0,1,2,\ldots,x} [V_1{}^*(y)+V_2 (x-y)]$ with $y_2{}^*(x)$ being the y that maximizes $V_2{}^*(x)$.

As a specific example for x = 7, then

$$V_2{}^*(7) = \text{Max}_{y=0,1,2,\ldots,7} [V_1{}^*(y)+V_2 (7-y)]$$
$$= \text{Max} [V_1{}^*(0)+V_2 (7-0), \quad V_1{}^*(1)+V_2 (7-1), \quad V_1{}^*(2)+V_2 (7-2),$$
$$V_1{}^*(3)+V_2 (7-3), \quad V_1{}^*(4)+V_2 (7-4), \quad V_1{}^*(5)+V_2 (7-5),$$
$$V_1{}^*(6)+V_2 (7-6), \quad V_1{}^*(7)+V_2 (7-7)]$$
$$= \text{Max} [0+143, 0+126, 39+106, 63+79, 78+44, 86+0, 92+0, 95+0]$$
$$= \text{Max} [143, 126, 145, 142, 122, 86, 92, 95] = 145$$
$$y_2{}^*(7) = 2$$
$$x_2{}^*(7) = 7-y_2{}^*(7) = 7-2 = 5$$

Table 6.2. Optimal Allocation for Technologies 1–3

Amount Available	Optimal Value Function for Technologies 1&2	Value Function of Technology 3	Optimal Value Function for Technologies 1,2&3	Optimal Allocation to Technology 1&2	Optimal Allocation to Technology 3
x	$V_2^*(x)$	$V_3(x)$	$V_3^*(x)$	$y_3^*(x)$	x_3^*
0	0	0			
1	0	**20**			
2	39	39			
3	63	46			
4	79	48			
5	106	49			
6	126	50			
7	145	50			
8	169	50			
9	189	50			
10	206	50	**209**	**9**	**1**

In Table 6.2, since Technology 3 is the last technology to be considered, we do not need to construct the entire optimal value function $V_3^*(x)$ for Technologies 1,2, and 3. Instead, we need only find $V_3^*(10)$. In this case,

$$V_3^*(10) = \text{Max}_{y=0,1,2,\dots,10}\ [V_2^*(y)+V_3(10-y)]$$

$$= \text{Max}\ [V_2^*(0)+V_3(10-0),\quad V_2^*(1)+V_3(10-1),\quad V_2^*(2)+V_3(10-2),$$
$$V_2^*(3)+V_3(10-3),\quad V_2^*(4)+V_3(10-4),\quad V_2^*(5)+V_3(10-5),$$
$$V_2^*(6)+V_3(10-6),\quad V_2^*(7)+V_3(10-7),\quad V_2^*(8)+V_3(10-8),$$
$$V_2^*(9)+V_3(10-9),\quad V_2^*(10)+V_3(10-10)]$$

$$= \text{Max}\ [\ 0+50,\quad 0+50,\quad 39+50,\quad 63+50,\quad 79+50,\ 106+49,\ 126+48,$$
$$145+46,\ 169+39,\ 189+20,\ 206+0]$$

$$= \text{Max}\ [50, 50, 89, 113, 129, 155, 174, 191, 208, 209, 206] = 209$$
$$y_3^*(10) = 9$$
$$x_3^*(10) = 10-y_3^*(10) = 10-9 = 1$$

Thus, the optimal value for allocating $10 million to the three technologies is $209 million obtained by allocating $9 million to Technologies 1 and 2 and $1 million to Technology 3. Then using Table 6.1, we find that the $9 million for Technologies 1 and 2 is best allocated by allocating $3 million to Technology 1 and $6 million to Technology 2. These best allocations are shown as bold-faced figures in the two tables.

6.7.4. Application to flexible manufacturing plant production mix[12]

The Dynamic Programming principle can also be applied to determine the optimal production mix for maximizing the total profit of a flexible manufacturing plant that can produce several types of products with various unit profits, plant capacity requirements, and demand quantities. We will illustrate this application with the following example.

A flexible manufacturing plant can be configured to produce batches of products 1, 2, and 3, with the unit profit and percentage of plant capacity required for producing each batch and the number of batches demanded for each of these products given below.

Product	Profit/Batch ($1000)	% Plant Capacity Required/Batch	Batches Demanded
i	P_i	R_i	D_i
1	35	50	1
2	8	10	2
3	20	33	3

To apply the Dynamic Programming principle, we will use the above information to first construct the value functions of these products from the following procedure:

Let x, P_i, R_i, and D_i be respectively the percentage of plant capacity available to, the profit per batch of, the percentage plant capacity required per batch of, and the batches demanded of Product i. Then

$$V_i(x) = \text{Int}(x/R_i)P_i \text{ if } x \le D_i R_i \text{ because a batch of item i requires } R_i \% \text{ capacity}$$
$$= D_i P_i \qquad \text{ if } x > D_i R_i \text{ because no more item i is available}$$

where Int (ratio) is the integer part of the ratio.

By this procedure, we have for example:

$$V_1(40) = \text{Int}(40/50)(35) = 0(35) = 0$$
$$V_1(60) = \qquad\qquad 1(35) = 35, \text{ because } x = 60 > D_1 R_1 = 1(50) = 50$$
$$V_2(30) = \qquad\qquad 2(8) = 16, \text{ because } x = 30 > D_2 R_2 = 2(10) = 20$$
$$V_3(70) = \text{Int}(70/33)(20) = 2(20) = 40$$

The complete set of value functions are given below.

---

[12] The simplified algorithm for this application is based on original work by the author.

% Plant Capacity	Value Function of Product i		
Available x	$V_1(x)$	$V_2(x)$	$V_3(x)$.
0	0	0	0
10	0	8	0
20	0	16	0
30	0	16	0
40	0	16	20
50	35	16	20
60	35	16	20
70	35	16	40
80	35	16	40
90	35	16	40
100	35	16	60

Applying the Dynamic Programming approach, we will then find the optimal production mix is: 1 batch of Product 1 using 50% of the capacity, 1 batch of Product 2 using 10% of the capacity, and 1 batch of Product 3 using 33% of the capacity for a total profit of $63,000 using a total of 93% of the capacity.

6.7.5. Major advantages and disadvantages

The Dynamic Programming approach is very efficient, particularly for large scale technology portfolio allocation decisions. The approach is theoretically elegant while it imposes no requirement on the form of the value functions as long as these functions are additive; i.e., the total value is the sum of the individual values. Furthermore, it can be applied to find not only the optimal resource allocation in technology portfolio decisions, but also the shortest route in a network and the best way to pack various items of different sizes, weights, unit values, and availabilities in a given total available space.

The major obstacle in applying this approach actually lies in the development of credible value functions. In a minor way, some textbooks have described the algorithm in highly convoluted ways that are difficult for practitioners to understand. As a result, this approach has not received the wide application it deserves.

6.8. References

Analytic Hierarchy Process

Saaty, T., *Decision Making for Leaders*, RWS Publications, 1999/2000 revised edition.

Saaty, T., *Multicriteria Decision Making: The Analytic Hierarchy Process*, RWS Publications, extended edition 1990.

Generalized LaGrange Multiplier Method

Everett, H., III, "Generalized Lagrange Multiplier Method for Solving Problems of Optimum Allocation of Resources," *Operations Research*, Vol. 11, pp. 399–417, 1963.

Linear Programming

Dantzig, G., *Linear Programming and Extensions*, Princeton University Press, 1998

Branch-and-Bound Approach

Lawler, E., and Wood, D., Branch-and-Bound Methods: A Survey," *Operations Research,* Vol. 14, pp. 699–719, 1966

Dynamic Programming

Bellman, R., *Dynamic Programming*, Dover, 2003

Software: Software for dynamic programming application has been developed by Gopal Bhat and is available from the author at www.starstrategygroup.com.

6.9. Exercises

Problem 6.1

For the computer purchase exercise in Problem 1.4, apply both the Multiple Rating Evaluation and the Analytic Hierarchy Process to determine your best choice and compare and discuss the differences of the results.

Problem 6.2

A pharmaceutical plant manager needs to minimize the cost of producing a type of diet pill from two ingredients, A and B, with respective unit costs of $3 and $5 per gram. Each gram of ingredient A contains 300 mg of multivitamins and 500 mg of minerals. Each gram of ingredient B contains 600 mg of multivitamins and 200 mg of minerals. Each diet pill weighs 1 gram and is required to have at least 360 mg of multivitamins and 300 mg of minerals. Use Linear Programming to determine the least cost combination of ingredients A and B in each diet pill. What are the respective shadow prices of the requirements for the multivitamins and the minerals?

Problem 6.3

A decision maker needs to assign 4 staff members A,B,C,D to be managers for 4 technology projects 1,2,3,4, with the requirement of one staff member for each project and one project per staff member. The costs in thousands of dollars for assigning various staff member to different project are summarized in the matrix below. Use the branch-and-bound approach to find the assignment that minimizes the total cost.

Costs of assigning staff members to technology projects

Project	1	2	3	4
Staff				
A	85	70	60	10
B	6	15	90	76
C	50	80	5	75
D	75	84	82	25

Problem 6.4

Plot the value functions in the example in Section 6.7.3 and apply the Generalized LaGrange Multiplier method to determine the optimal allocation of $10 million to the three projects. Compare the result with that obtained from the Dynamic Programming approach and discuss if there is any difference.

Problem 6.5

A flexible manufacturing plant can be configured to manufacture batches of products 1, 2, 3, and 4. The unit profit and percentage of plant capacity required for producing each batch together with the number of batches demanded for each of these products are summarized in the table below.

Product	Profit/Batch ($1000)	% Plant Capacity Required/Batch	Batches Demanded
1	8	10	2
2	13	20	3
3	20	30	2
4	30	40	1

Use the Dynamic Programming approach to find the product mix that maximizes the total profit.

7   Find the Optimal Technology Portfolio:
Decision Under Uncertainty

As technology development or application evolves over time and the impact on the decision maker's values takes place in the future, the technology portfolio planning and management process is fraught with uncertainty. In Chapter 6, we have used an aggregated expected value of the uncertain outcomes as a deterministic equivalent for the relationships between alternatives and values in the process. However, the deterministic equivalent is in many situations too aggregated for proper analysis, and the uncertainty effects need to be considered in greater detail. In this chapter, we will present major methods that will assist the decision maker in finding the best technology portfolio by explicitly analyzing the effects of uncertainty.

7.1. Decision Tree Analysis

A well-established approach for modeling the decision making process over time is the decision tree. By using the tree structure, we can systematically analyze the effects of uncertainty on the outcomes of different alternatives and determine the optimal choice for technology planning decisions.

7.1.1. Decision tree structure

This method assumes that the decision process is interlaced with two types of time points:

1. Those time points where the decision maker has a choice about what will happen next, which are called Decision or Choice Nodes. They are conventionally represented by squares with alternatives as emanating branches.
2. Those time points where the decision maker has little control and must let future events unfold in a probabilistic manner, which are called Probability or Chance Nodes. They are conventionally represented by circles with outcomes as emanating branches.

A decision tree is then the tree-shaped diagram formed by a sequence of choice nodes, alternative branches, chance nodes, and outcome branches interlaced over time.

A decision tree generally starts with a single choice node and ends with a set of chance nodes with final outcomes. However, in between, there is no set sequence of these nodes, as a decision node can be followed by another decision node, and a probability node can be followed by another probability node. The time point of a decision node is often called a stage.

For a major technology portfolio decision process over time, the question of when the process or its associated decision tree should end is generally taken care by future discounting, i.e., the values of future outcomes are discounted over time and eventually diminish to zero. For technology portfolio planning and management, the process generally considers a 5- to 10-year time horizon.

As shown in Figure 7.1, the utility and risk attitude analysis discussed in Chapter 2 can be easily represented by a single-stage decision tree.

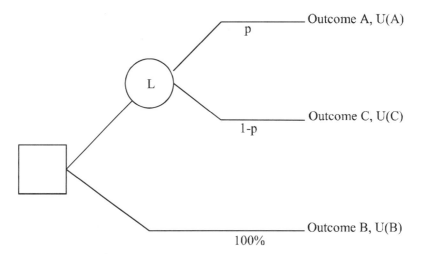

Figure 7.1. A Single-Stage Decision Tree for Utility Determination

As another example, a simplified technology planning decision is shown as a single-stage decision tree in Figure 7.2. In this example, the decision maker faces the choice of investing $5 million in one of two alternative technology projects: (1) developing a new technology that if successful will yield a profit of $15 million (net profit $10 million), but if not successful will result in $0 profit (net loss $5 million), or; (2) improving an existing technology that is sure to yield a profit of $7 million (net profit $2 million). Here, the choices are the technology projects, which is under the control by the decision maker. However, once a project has been chosen, whether the technology will eventually achieve economic feasibility, implementation effectiveness, and market profitability is often beyond the direct control of the decision maker, hence it is a chance event.

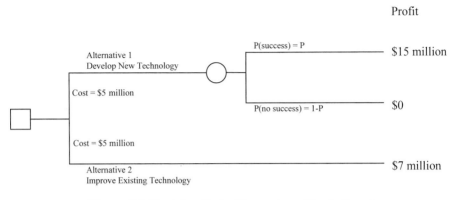

Figure 7.2. Decision Under Uncertainty (Single-Stage)

7.1.2. Analysis procedure

Once the decision tree is formulated, it can be analyzed through the following steps:

1. Estimate probabilities for the outcome branches of each chance node
2. Assign values to the outcome branches of each chance node at the end of the tree
3. Start at the end of the tree and go backwards:
   (a) Compute the expected value of each chance node, which is the expected value of its associated alternative branch
   (b) Select the alternative branch with the highest expected value and eliminate other branches at each choice node
4. Stop when the initial choice node has been reached.

7.1.2.1. Probability estimation

As discussed in the Appendix, the probabilities to be assigned at the outcome branches of each chance node can be obtained through a combination of classical analysis based on equal likeliness of elementary events, statistical analysis of empirical data, and subjective judgments. Further, if an alternative involves new information acquisition, Bayes Theorem can be used to revise these probabilities based on the new information.

It is interesting to note that a single-stage decision-tree approach can be applied to quantify the subjective judgments. As an example, we will estimate the probability of success, P, for the new technology in Figure 7.2. In this application, the decision maker is asked to choose between two alternative lotteries, L1 and L2, as shown in Figure 7.3.

Lottery L1 has two outcomes, one is the most desirable event A and the other is the least desirable event B. The probabilities of occurrence of these events are, respectively, the probabilities of success, P, and of failure, 1–P, of the new technology of which we would like to ask the decision maker to estimate based on subjective judgment.

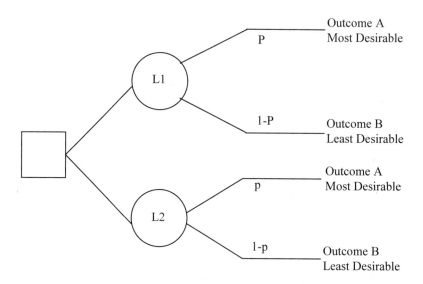

Figure 7.3. A Single-Stage Decision Tree for Subjective Probability Estimation

Lottery L2 also has the same two outcomes, A and B, with respective probabilities of occurrence of p and 1–p which can be varied.

Clearly, if p is set at 0, a decision maker will choose L1, and if p is set at 1, then the decision maker will choose L2. As p increases from 0 to 1, by Axiom (5) of the utility theory in Chapter 2, there exists p^* at which the decision maker becomes indifferent between L1 and L2. Then $p^* = P$, the decision maker's subjective judgment of the probability of success for the new technology.

7.1.2.2. Illustrative example for decision under uncertainty

As an illustrative example, Figure 7.4 shows the analysis results of the decision tree in Figure 7.2. We assume that the decision maker is risk neutral so the value is simply the monetary value. Furthermore, the decision maker has used the lottery approach to estimate the subjective judgment for the probability of success, P, of the

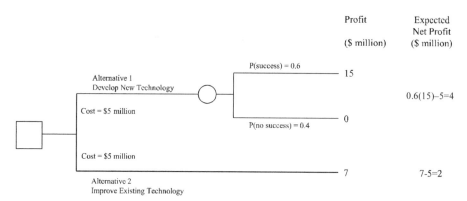

Figure 7.4. Decision Under Uncertainty (Single-Stage)

new technology by feeling indifferent between these two lotteries when $P = p^* = 0.6$. The analysis shows that developing new technology has a higher expected net profit and is thus the optimal choice.

7.1.3. Single-stage decision tree analysis—breakeven analysis

Another application of single-stage decision tree analysis is the breakeven analysis.

7.1.3.1. Decision tree structure

In this application, as shown by the generic decision tree in Figure 7.5, the decision process involves a series of alternatives, A_i, i=1, 2,...,m, where each alternative, A_i, has a range of probable outcomes, O_{ij}, j=1, 2, ...,n_i, with probabilities p_{ij} respectively. Thus, each alternative A_i will have expected value $V_i = \Sigma_j\, p_{ij} O_{ij}$. Clearly, the optimal alternative is the one with the largest expected value.

However, as a special case, if the expected values of these outcomes for the range of alternatives are the results of two opposing functions, with the value of one function increasing with the index i at a diminishing rate, the value of the other function decreases with i at a constant or increasing rate, then the optimal alternative will be the *breakeven* i at which the marginal increment from one function equals the marginal decrement from the other function.

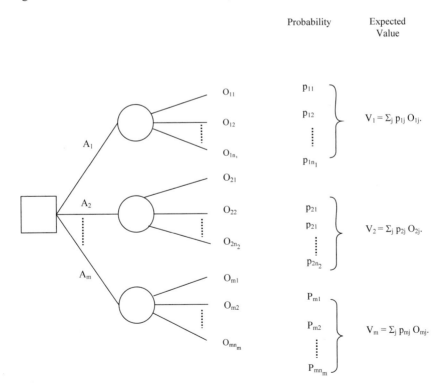

Figure 7.5. A Generic Decision Tree for Breakeven Analysis

7.1.3.2. Illustrative example

As a specific example, a manufacturer produces computers at the cost of $500 per unit and sells them for $1000 per unit. If a computer is produced but not sold, it has a salvage value of $100. Demand for computers from this manufacturer is estimated to be normally distributed with mean 100,000 units and standard deviation 10,000 units. What is the optimal number of units that the manufacturer should produce to maximize its expected net revenue?

The marginal profit of an sold computer is   MP=$1000–$500=$500
The marginal loss of an unsold computer is   ML=$500–$100=$400

If the manufacturer produces S units, then

Expected profit $= [\sum_{n=0}^{n=S} n\, P(Demand=n) + S\, P(Demand >S)]\, MP$
Expected loss $= [\sum_{n=0}^{n=S} (S\text{-}n)\, P(Demand=n)]\, ML$

Now, if the manufacturer produces S+1 units, then

Expected profit $= [\sum_{n=0}^{n=S} n\, P(Demand=n) + (S+1)\, P(Demand >S+1)]\, MP$
$= [\sum_{n=0}^{n=S} n\, P(Demand=n) + (S+1)\, P(Demand \geq S+1)]\, MP$
$= [\sum_{n=0}^{n=S} n\, P(Demand=n) + (S+1)\, P(Demand >S)]\, MP$
Expected loss $= [\sum_{n=0}^{n=S+1} (S+1\text{–}n)\, P(Demand=n)]\, ML$
$= [\sum_{n=0}^{n=S} (S+1\text{–}n)\, P(Demand=n)]\, ML$

Take the differences between the two expected profits and the two expected losses, then

Expected marginal profit from an additional unit
$= P(Demand>S)MP$ which decreases with S
Expected marginal loss from an additional unit
$= P(Demand\leq S)]ML$ which increases with S.

Thus, the S that maximizes the net expected profit is the breakeven amount at which

$P(Demand>S)MP = P(Demand\leq S)]ML$

Or equivalently,

$[1–P(Demand\leq S)]MP = P(Demand\leq S)]ML$

or when $P(Demand\leq S) = MP/(ML+MP)=500/(400+500)=0.556$

With normally distributed demand, and $P(Z<0.15)=0.556$, the optimal choice is to produce $100000+0.15(10000)=101500$ units, where Z is the standard normal random variable.

7.1.4. Two-stage decision tree analysis—expected value of information

A major application for a two-stage decision tree analysis is to estimate the expected value of information. In decision making, it is often possible to acquire further information about an uncertain chance event. For example, for the technology portfolio decision in Figure 7.2, further information about the new technology's economic feasibility may be acquired through prototype development and information about market profitability may be acquired through market research. However, this information acquisition generally carries a significant cost. Thus, whether to acquire information or not is again a decision that should be determined by whether the expected value of the information to be acquired exceeds the cost for the information.

7.1.4.1. Decision tree structure

For the example in Figure 7.2, the decision maker is considering hiring a technology forecasting firm to predict whether the new technology would be successful in the market. The two-stage decision tree for whether to acquire the forecast is given in Figure 7.6

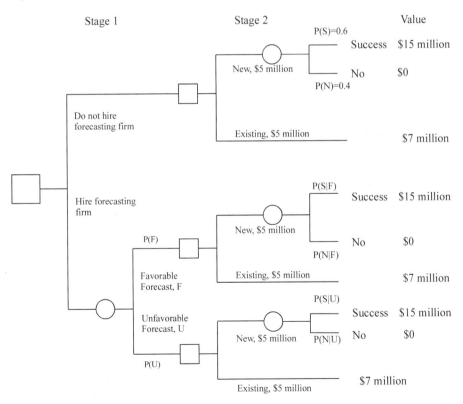

Figure 7.6. Two-Stage Decision Under Uncertainty (General Case)

7.1.4.2. Estimation of the conditional and posterior probabilities

Since the information acquired will affect the decision maker's judgment of the probability of success for the new technology, we need to first estimate the reliability of the information to be acquired based on the historical reliability of the information provider. Specifically for this example, let F and U denote respectively favorable and unfavorable forecasts, and S and N denote respectively whether a technology will truly be successful or not. The technology forecasting company has provided the following historical forecasting records for technologies similar to the new technology under consideration below:

> 90% of those technologies that were proven successful in the market, the company had given a favorable forecast, i.e., $P(F|S)=0.9$
> However, for the 20% of those technologies that have failed in the market, the company had given a favorable forecast, i.e., $P(F|N)=0.2$

For the decision tree analysis, the decision maker then needs to use the Bayes Theorem to estimate $P(S|F)$, $P(N|F)$, $P(S|U)$, and $P(N|U)$, i.e, the respective posterior probabilities that the new technology will be truly successful or not in the case when the forecast is favorable and in the case when the forecast is unfavorable. In addition, since the decision maker has not yet chosen to hire the forecasting firm, the decision maker needs to estimate a priori $P(F)$ and $P(U)$, i.e., the probabilities that the forecast would be favorable or unfavorable.

By Bayes Theorem, we have
$$P(S|F) = P(F|S)P(S)/[P(F|S)P(S)+P(F|N)P(N)]$$
$$= 0.9(0.6)/[0.9(0.6)+0.2(0.4)]=0.54/0.62 = 0.87$$
$$P(S|U) = P(U|S)P(U)/[P(U|S)P(S)+P(U|N)P(N)]$$
$$= 0.1(0.6)/[0.1(0.6)+0.8(0.4)]=0.06/0.38 = 0.16$$
$$P(N|F) = 1-P(S|F) = 0.13; \qquad P(N|U) = 1-P(S|U) = 0.84$$
$$P(F) = 0.62; \qquad\qquad\qquad P(U) = 0.38$$

7.1.4.3. Expected value of information

Based on these probabilities, the risk-neutral decision maker would choose the new technology if the forecast is favorable because it has a higher expected profit, and choose the existing technology if the forecast is unfavorable because the new technology would have a lower expected profit. As a result, the expected net profit with the forecast information is

$$0.62[0.87(15-5)+0.13(0-5)]+0.38(7-5) = \$5.76 \text{ million}$$

In comparison, the expected net profit without the forecast is $4 million as shown in Figure 7.4. Thus, we have

Expected value of the forecast information = $5.76-$4.0 = $1.76 million and a risk-neutral decision maker will hire the forecasting firm if it charges less than $1.76 million.

7.1.4.4. Expected value of perfect information

For the special case of perfect information, i.e., $P(F|S)=P(U|N)=1$, we will have
$P(S|F)=1$, $P(N|F)=0$;
$P(S|U)=0$, $P(N|U)=1$
$P(F) = 0.6$, $P(U)=0.4$

Based on this perfect information, the decision maker would choose the new technology if the forecast is favorable and the existing technology if it is unfavorable, and the expected value with the forecast is

$0.6(15–5)+0.4(7–5) = \$6.8$ million

Thus, the expected value of perfect information $=\$6.8–\$4=\$2.8$ million

Clearly, the perfect information has the greatest expected value. However, from the decision tree, we can see that if every outcome in the case with the information results in the same optimal choice as in the case of no information, then the information has no value, even when the information is perfect. To illustrate this point with a specific example, if the net profit from improving the existing technology is $10 million, then even if the forecast for the new technology is perfect, a risk-neutral decision maker will choose to improve the existing technology.

7.1.5. Major advantages and disadvantages

The decision tree structure provides a reasonable approximation of the time evolving and uncertain nature of the decision making process. Furthermore, the analysis procedure is systematic and incorporates the effects of uncertainty properly.

However, in real world applications, the decision tree structure can be very large, unwieldy, and difficult to manage.

More importantly, the structure assumes that *all* future potential alternatives and outcome probabilities are *known at the present*. In reality, many, if not most of the future possibilities are unknown at the present, only to be discovered or understood when the future unfolds. (For example, it is often impossible for technology developers and investors to have a full understanding, let alone detailed probability estimation, of the possible advances in aerospace, telecommunications, and computer technologies at the onset of these technologies.) Thus, the decision tree approach cannot effectively reflect the adaptive (i.e., learning by doing) or feedback nature of sequential decision making where alternatives and outcomes are not fully known at the beginning but gradually revealed over time. A possible remedy to this limitation would be to regularly review and revise the decision tree as new information is acquired.

7.2. Portfolio Diversification

An important tool in technology portfolio planning and management is portfolio diversification. The concepts discussed in this section borrow heavily from financial portfolio diversification, which has been well-established since the seminal work of Harry Markowitz in 1952.

7.2.1. Portfolio value and risk

In technology development or application, there are many potential outcomes with positive values, such as profit, contribution to environmental quality, and prestige. On the other hand, there are also many potential outcomes with negative values, such as physical failures, financial losses, environmental or safety disasters, and image damages. A decision maker would naturally like to reduce the chance or risk of encountering outcomes with negative values. Using the value assessment and probability estimation techniques in Chapter 2 and the Appendix, respectively, it is possible for a decision maker to develop a probability distribution of these values. Thus, from a probability distribution point of view, one approach the decision maker can use to reduce this chance is to concentrate on alternatives with high expected values but low volatilities, thus lower probabilities of having outcomes with negative values. Another approach is to develop a portfolio of complementary technologies, so that the combined volatility is reduced without great loss in the combined expected value. These two concepts are the essence of this section.

Since risk is the chance of encountering outcomes with negative values, it can be represented by the volatility in the values, which in turn is measured by the standard deviation of probability distribution of the value of a portfolio. For a typical set of portfolios, the respective expected values, μ's, and risks, as represented by the standard deviations, σ's, of the value probability distributions of these portfolios are shown by the umbrella shaped area in Figure 7.7. The unusual shape of the area is the result of portfolio diversification as to be discussed below.

7.2.2. Benefit of diversification

For two technologies A and B with expected values μ_A and μ_B and standard deviations σ_A and σ_B for their respective value probability distributions, if their values are correlated with correlation coefficient ρ, then based on probability theory, a diversified portfolio with p proportion of A and 1–p proportion of B will have

$$\text{Expected value} = \mu = p\mu_A + (1-p)\mu_B$$
$$\text{Standard deviation} = \sigma = \sqrt{[p^2\sigma_A^2 + (1-p)^2\sigma_B^2 + 2\rho\, p(1-p)\sigma_A\sigma_B]}$$

Thus, for a given value of ρ, the portfolio expected value and standard deviation will vary for different values of p. The μ's and σ's of various portfolios of two technologies with given expected values and standard deviations but different correlation coefficient ρ and portfolio proportion p are shown in Table 7.1 and Figure 7.8.

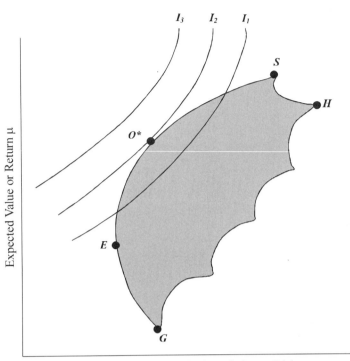

Figure 7.7. Expected Values and Risks of Technology Portfolios

Table 7.1. Portfolio Risk as Functions of Portfolio Composition and Correlation for Two Technology Investments with $\mu_A = 0.2$, $\sigma_A = 0.1$ and $\mu_B = 0.1$, $\sigma_B = 0.05$

	Proportion of A in the Portfolio, p					
	0.000	*0.250*	*0.3333*	*0.500*	*0.750*	*1.000*
Portfolio Return, μ	0.100	0.125	0.1333	0.150	0.175	0.200
Correlation ρ	Portfolio Risk, σ					
−1.0	0.050	0.0125	0.0000	0.025	0.0625	0.100
−0.5	0.050	0.0331	0.0333	0.0433	0.0696	0.100
0.0	0.050	0.0451	0.0471	0.0559	0.0760	0.100
0.5	0.050	0.0545	0.0577	0.0661	0.0820	0.100
1.0	0.050	0.0625	0.0667	0.0750	0.0875	0.100

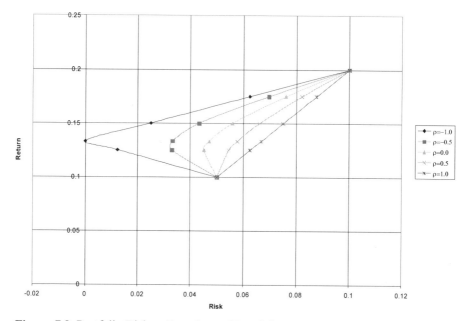

Figure 7.8. Portfolio Risk as Functions of Portfolio Composition and Correlation for Two Technology Investments with $\mu_A = 0.2$, $\sigma_A = 0.1$ and $\mu_B = 0.1$, $\sigma_B = 0.05$

It is interesting to observe from Table 7.1 and Figure 7.8 that when the correlation coefficient of the two technologies is negative, the diversified portfolio can have lower risks than that of either technology without significant reductions in the combined expected value. In the special case of $\rho = -1$, if $p = \sigma_B/(\sigma_A + \sigma_B)$, then the portfolio standard deviation will be 0. In other words, with such a diversified portfolio, we can totally eliminate the risk or chance of outcomes with negative values and have the combined expected value of $p\mu_A + (1-p)\mu_B$ with certainty.

This analysis can also be generalized to portfolios with three or more investments. As a illustrative example, for three technology investments, X, Y, and Z, with respective expected values, standard deviations, and correlation coefficients given below, the (σ, μ) or risk-return curves of various portfolios of these investments are shown in Figure 7.9. The envelope of these curves forms the umbrella shape shown in Figure 7.7.

Investment	Expected Value (Return)	Standard Deviation (Risk)	Correlation Coefficient
X	0.2	0.1	$\rho_{xy} = -0.5$
Y	0.1	0.05	$\rho_{xz} = -0.2$
Z	0.15	0.06	$\rho_{yz} = 0.33$

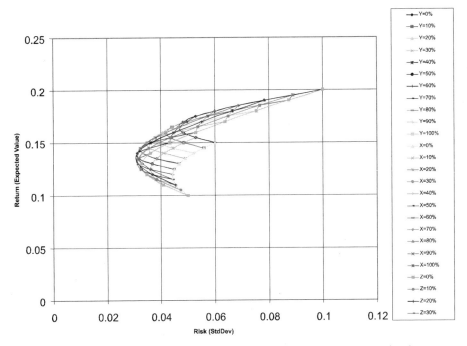

Figure 7.9. Risk-Return Curves of Various Portfolios of Three Technology
Investments

It can also be observed from Figures 7.8 and 7.9 that the risk-return curve for
given correlation coefficients is concave, in that the linear combination of any two
points on the curve lies either on or to the right side of the curve. Since any two
portfolios can be combined into a larger portfolio, the envelope of the (σ, μ)'s of all
possible portfolios will also be concave as shown by the line SOEG in Figure 7.7.

7.2.3. Efficient frontier, iso-preference curve, and optimal portfolio

In Figure 7.7, let S be the (σ, μ) with the highest expected value and E be the (σ, μ)
with the smallest standard deviation among all portfolios. Then those portfolios that
lie on the curve SE form the Efficient Frontier, in that each portfolio on the curve has
the smallest standard deviation or risk among all portfolios with the same expected
value and the highest expected value among all portfolios with the same standard
deviation or risk.

On the other hand, using value assessment techniques in Chapter 2, a decision
maker can determine the preference tradeoffs between expected return and standard
deviation of the return (risk). Specifically, for a portfolio with a given expected
return and standard deviation (risk), there are many other portfolios either with
higher expected returns but also higher standard deviations (risks); or with lower
standard deviations (risks) but also lower expected returns that the decision maker
will prefer equally. The (σ, μ)s of these equally preferred portfolios form iso-
preference curves, indicated, for example, by I_1, I_2, and I_3 in Figure 7.7.

These iso-preference curves have the following properties:

1. They are necessarily parallel to one another, because otherwise the intersection of two curves would represent a (σ, μ) with two different degrees of preferences by the decision maker, a self-contradiction.
2. For two portfolios with equal expected values but different standard deviations or risks, a decision maker will always prefer the portfolio with the smaller standard deviation or risk. As a result, the iso-preference curve to the left of another iso-preference curve will have a higher degree of preference. In Figure 7.7, I_3 has a higher degree of preference than I_2, which in turn has a higher degree of preference than I_1.
3. By diversification, the linear combination of two equally preferred portfolios will yield a portfolio with equal or lower risks, depending on the correlation between the portfolios. Thus, for a risk neutral decision maker, the iso-preference curve is necessarily convex in that the linear combination of two points on the curve lies either on or to the left of the curve.

Given these properties, there will be an iso-preference curve that intersects the efficient frontier at a single tangent point, as shown as the point O in Figure 7.7. This is then the optimal portfolio for the particular decision maker.

7.2.4. Major advantages and disadvantages

As well-known intuitively, diversification reduces the risk of a portfolio. The analytical approach discussed in this section quantifies that benefit. Furthermore, it provides a systematic procedure for determining the optimal portfolio.

However, unlike financial portfolios, there is a paucity of available information about returns and risks for technology portfolios, especially for emerging technologies. As a result, the value probability distributions of technologies and correlation coefficient among these values tend to be judgmental and speculative, which in turn can render the analytical results not reliable or credible.

Nevertheless, such analysis can lend important insight to not only the need for diversification but also the general degree of diversification, i.e., the rough proportions of technologies in a portfolio that would be desirable.

7.3. Real Options Theory and Application

In the development or application of large, complex, and uncertain technologies, a decision maker often has a number of options, such as deferring investment to gain more information through a pilot study; and expanding, contracting, and abandoning a project as it progresses, that can reduce the risk of undesirable outcomes and increase the managerial flexibility to counter such outcomes. Having an option generally carries a cost, such as the cost of the pilot study and loss from abandoning a project. Thus, a key question would be how to estimate the value of an option. One approach is to use the decision analysis methodology discussed in Section 7.1. However, as will be discussed later in this section, there are certain shortcomings in using decision-tree analysis for financial analysis. Instead, an innovative and effective approach has been to apply the well-established financial options theory to technology options. Since technology, like real estate or a physical facility, is a real property asset, the theory underlying such applications is called Real Options Theory.

In the last two decades, there have been considerable research and applications of real options theory. This section will focus on the basic concepts and present a number of simple examples to illustrate how real options theory can be to used to provide flexibility and insight in technology portfolio planning and management decisions. Interested readers can obtain additional information from the sources listed in the reference section.

7.3.1. Basic concepts of options theory

In financial options theory, option is the right, but not the obligation, to follow through on a business decision. There are basically two complementary types of options:

The Call option gives the option purchaser the right but not the obligation to buy from the option writer (usually a broker but also could be the owner of the asset) a specific amount of an asset of interest, including financial assets such as stocks and real assets such as raw materials, from the asset owner at a pre-agreed exercise or strike price for purchasing the asset, either at (as in the European options) or by (as in the American options) a prescribed expiration date of the option contract[13]. In this case, it is important to note that if the option holder exercises the right, then the Call option seller must fulfill the contract by selling the specific amount of the asset of interest at the pre-agreed strike price to the option holder.

Similarly, the Put option gives the option purchaser the right to sell to the option writer a specific amount of an asset of interest at a pre-agreed exercise or strike price either at or by the expiration date of the option contract. If the option holder exercises the right, the Put option seller must also fulfill the contract by purchasing the specific amount of the asset of interest at the pre-agreed strike price from the option holder.

At the time of exercising the option right, if the market price for the asset is higher than the strike price, then the Call option holder will gain a profit by buying the asset at an agreed lower strike price and selling it to the market at a higher price and be "in the money." On the other hand, if the market price is lower than the strike price, then the Call option holder can abandon the option and be "out of money," but not incur any additional loss.

The reverse is true for the Put option holder; i.e., if the market price is higher than the strike price, the Put option holder can abandon the option and be "out of money" but without additional loss, while if the market price is lower, the option holder would gain a profit by buying the asset from the market at a lower price and selling it to the option seller at a higher pre-agreed strike price and be "in the money."

---

[13] For simplicity, we will focus exclusively in this book on the European option scheme, i.e., the option can only be exercised at the time of contract expiration.

For the advantages provided by the option, the option purchaser needs to pay a price or premium. Specifically, for the Call option, the option purchaser needs to pay a premium to compensate the option seller the opportunity cost for not selling the asset to others even if the market price is higher. The Call option purchaser also should pay this premium for the privilege of hedging against the risk of potential increases in the asset price in the market as well as for the flexibility of not buying the asset if its strike price should become higher than the market price.

Similarly, for the Put option, the option purchaser needs to pay a premium to compensate the option seller for buying the asset at the pre-agreed price even if the market price is lower. The Put option purchaser also should pay this premium for the privilege of hedging against the risk of potential decreases in the asset price in the market as well as for the flexibility of not selling the asset if its strike price should become lower than the market price.

As an example of a Call option with technology application, a company is leasing a technology innovation from its owner that has a current market price of $200,000, with an option to buy the innovation for the strike price of $220,000 at the end of the lease. To obtain this option, the company needs to pay the innovation owner a premium to compensate for the opportunity cost to the owner for not selling the innovation to others or raising the price in the future. The company pays this premium also for the privilege of hedging against the risk of potential increases in the innovation price in the market at the end of the lease as well as for the flexibility of not buying the innovation if the value of the innovation falls below the strike price or a better innovation becomes available at the end of the lease.

7.3.2. Option valuation

The main factors affecting the value of an option under the European scheme are:

- K, the strike price of the underlying asset, or the agreed price for the Call option buyer to purchase or for the Put option buyer to sell the asset at the end of the option contract
- S, current price of the underlying asset
- Option contract expiration date
- Price volatility of the underlying asset
- r, risk-free interest rate

7.3.2.1. The Single Period Binomial Case

The simplest case for the valuation of an option under the European scheme is the single period binomial outcome case where there are only two possible outcomes for the market price of the underlying asset at the date of expiration. For option valuation for this case, we assume that the market price of the underlying asset, which is currently at S, will either go up by a factor u or go down by a factor d as follows:

Upward: price is uS with probability p, where $u>1$
Downward: price is dS with probability $1-p$, where $u>d>0$

Let R = 1+r with r being the risk-free interest rate from, for example, a government-insured bond that one can buy or sell. It is necessary that u>R>d, because if R>u, then there is no need for the option; on the other hand, if R<d, then one can create a riskless profitable arbitrage by borrowing a risk-free loan at interest rate r to purchase the option contract with assured greater return, which cannot exist in an efficient market. At the date of option contract expiration, we have for a Call option:

C_u = worth of the Call option when market price is uS = max (uS – K, 0)
C_d = worth of the Call option when market price is dS = max (dS – K, 0)

or for a Put option:

P_u = worth of the Put option when market price is uS = max (K-uS, 0)
P_d = worth of the Put option when market price is dS = max (K-dS, 0)

We can duplicate the above result by buying x dollars amount of the underlying asset and b dollars amount of the risk-free bond (or borrow a loan if b is negative), so that at the date of option contract expiration, this portfolio will be worth for a Call option:

$$ux + Rb = C_u$$
$$dx + Rb = C_d$$

or for a Put option

$$ux + Rb = P_u$$
$$dx + Rb = P_d$$

Solving the above equations, we find that for a Call option:

$$x = (C_u - C_d)/(u\text{-}d)$$
$$b = (C_u - ux)/R = (uC_d - d\, C_u)/[R(u\text{-}d)]$$

or for a Put option:

$$x = (P_u - P_d)/(u\text{-}d)$$
$$b = (P_u - ux)/R = (uP_d - d\, P_u)/[R(u\text{-}d)]$$

Since x+b has the same outcomes as the option, we have

C = the value of the Call option = x+b = $[(R\text{-}d)\, C_u + (u\text{-}R)\, C_d]/[R(u\text{-}d)]$ (7.1)
or   P = the value of the Put option = x+b = $[(R\text{-}d)\, P_u + (u\text{-}R)\, P_d]/[R(u\text{-}d)]$ (7.2)

If we let q = (R-d)/(u-d), which is called the risk-neutral ratio, then

$$C = [qC_u + (1-q)C_d]/R$$
or $$P = [qP_u + (1-q)P_d]/R$$

It is interesting to see that the probability p does not enter into the valuation process. The reason is that p has already been factored into the future prices, uS and dS, through the risk-adjusted discount factor. In other words, the investor should have already taken into account the risks inherent in the future prices, so that the present value of the expected future price = [puS + (1–p)dS] / (1+risk-adjusted discount factor) = S.

The advantage of the use of the risk-neutral probability q instead of p is that the risk-free interest rate r becomes the discount factor for the present value computations. Thus, we will not need to adjust the discount factor in a different situation. This is a major advantage of real options analysis over decision tree analysis in option valuation.

Example
For the lease option of a technology innovation with a current price of $200,000, and a strike price of $220,000 at the end of a one-year lease, assume that at the end of the lease, the market price for the innovation may be 25% higher or lower, i.e., u=1.25 and d=0.75, and that the risk-free interest rate is 3% per year, i.e., r=0.03. Then we have

$$C_u = \max (\$250{,}000 - \$220{,}000, 0) = \$30{,}000$$
$$C_d = \max (\$150{,}000 - \$220{,}000, 0) = 0$$

Now, if we purchase x dollar amount share of the technology innovation (in reality, we would negotiate with the owner to purchase partial ownership of the technology innovation at the end of the lease) and b dollar amount of the risk-free bond at 3% annual interest, then at the end of the year we will have

$$1.25x + 1.03b = \$30{,}000$$
$$0.75x + 1.03b = 0$$

Solving the above equations yields x = $60,000 and b = –$43,689, which means that we will purchase (60,000/100,000), or 30% share of the technology innovation and borrow $43,689 of risk-free loan at 3% annual interest. As a result,

C = the value of the Call option = x+b = $16,311

or if we apply Equation (1), we have equivalently,

$$C = [(1.03-0.75) C_u + (1.25-1.03) C_d]/[1.03(1.25-0.75)] = (0.28)(\$30{,}000)/0.515$$
$$=\$16{,}311$$

In summary, the value for the lease option to the purchaser is $16,311. If the seller asks for less, the purchaser should enter into the option contract; otherwise, the purchaser should forgo the option. Furthermore, it is easy to see that the value of the Call option increases with decreasing strike price and increasing price volatility or u-d.

It is important to note that, although the formulas for Call and Put options are very similar, buying a Call option is totally different from buying a Put option in terms of rights and obligations; thus, C and P are unrelated and not equal to each other.

7.3.2.2. Two-Period and Multiple-Period Binomial Cases

Two-period
We can extend the valuation process for the single period binomial case to a two-period binomial case. In this case, there are two possible outcomes in each period. For simplicity, let the outcomes be similar in each case; i.e., at the end of each period, the price of the underlying asset will be either u or d times the price at the beginning of the period. Thus, with S being the current price of the underlying asset, at the end of Period 1, the market price for the underlying asset will have two possible values: uS and dS. Then, at end of Period 2, the market price for the asset will have the following possibilities:

If it is uS at the end of Period 1, then uuS or duS at the end of Period 2
If it is dS at the end of Period 1, then udS or ddS at the end of Period 2

Now, at the end of Period 2, we have:

C_{uu} = worth of the Call option when market price is uuS = max (uuS-K, 0)
$C_{du} = C_{ud}$ =worth of the Call option when market price is duS =
    max (duS-K, 0)
C_{dd} = worth of the Call option when market price is ddS = max (ddS-K, 0)
or P_{uu} = worth of the Put option when market price is uuS = max (K-uuS, 0)
$P_{du} = P_{ud}$ =worth of the Put option when market price is duS =
    max (K-duS, 0)
P_{dd} = worth of the Put option when market price is ddS = max (K-ddS, 0)

Following the same process as the single period binomial case, we have

C_u = the value of the Call option at the beginning of Period 2 if market
    price at the end of Period 1 is uS = $[(R-d) C_{uu} +(u-R) C_{du}]/[R(u-d)]$
C_d = the value of the Call option at the beginning of Period 2 if market
    price at the end of Period 1 is dS = $[(R-d) C_{ud} +(u-R) C_{dd}]/[R(u-d)]$
or P_u = the value of the Put option at the beginning of Period 2 if market
    price at the end of Period 1 is uS = $[(R-d) P_{uu} +(u-R) P_{du}]/[R(u-d)]$
P_d = the value of the Put option at the beginning of Period 2 if market
    price at the end of Period 1 is dS = $[(R-d) P_{ud} +(u-R) P_{dd}]/[R(u-d)]$

With q = (R-d)/(u-d) being the risk-neutral probability of upside price, then

$C_u = [qC_{uu} +(1-q)C_{du}]/R, C_d = [qC_{ud} +(1-q)C_{dd}]/R$
or $P_u = [qP_{uu} +(1-q)P_{du}]/R, P_d = [qP_{ud} +(1-q)P_{dd}]/R$

Through one more backward iteration, we have

$$
\begin{aligned}
C &= \text{the value of the Call option at Period 1} \\
&= [(R\text{-}d)\,C_u + (u\text{-}R)\,C_d]/[R(u\text{-}d)] \\
&= [qC_u + (1\text{-}q)C_d]/R \\
&= \{q[qC_{uu} + (1\text{-}q)C_{du}] + (1\text{-}q)\,[qC_{ud} + (1\text{-}q)C_{dd}]\}/R^2 \\
&= [q^2 C_{uu} + 2q(1\text{-}q)C_{ud} + (1\text{-}q)^2 C_{dd}]/R^2 \qquad (7.3)
\end{aligned}
$$

or $\quad P$ = the value of the Put option at Period 1

$$
= [q^2 P_{uu} + 2q(1\text{-}q)P_{ud} + (1\text{-}q)^2 P_{dd}]\}/R^2 \qquad (7.4)
$$

Multiple-Period
This process can be extended to the binomial case for any number of periods. By induction, we can see that the C can be expressed as a function of the various possible worths of the Call option at the end of the nth period. The coefficients of the possible worths of the Call option at the end of the nth period follows a binomial series of the parameter q, which is defined as the risk neutral probability.

Specifically, for n periods, let $u(x)d(n\text{-}x)$ symbolize x number of u's followed by n-x number of d's, then we have

$$
C = \sum\nolimits_{x=0,\dots,n} \cdot \left[\binom{n}{x} q^x (1-q)^{n-x} \cdot C_{u(x)d(n-x)} \right] / R^n \qquad (7.5)
$$

where $\binom{n}{x}$ is the number of combinations of x objects selected out of a total of n objects.

$$
C_{u(x)d(n-x)} = \max\,[u(x)d(n\text{-}x)S\text{-}K,\,0]
$$

Similarly for the Put option with n periods,

$$
C = \sum\nolimits_{x=0,\dots,n} \cdot \left[\binom{n}{x} q^x (1-q)^{n-x} \cdot P_{u(x)d(n-x)} \right] / R^n \qquad (7.6)
$$

where $\binom{n}{x}$ is the number of combinations of x objects selected out of a total of n objects.

$$
P_{u(x)d(n-x)} = \max\,[K\text{-}u(x)d(n\text{-}x)S,\,0]
$$

Examples
(1) Again, use the example of the lease option of a technology innovation with the same parameter values as in Section 7.3.2.1, except that the lease is now two years long and the purchaser of the option can exercise the option at the end of each year.

In this example, we have

$R = 1.03$

$q = (1.03–0.75)/(1.25–0.75) = 0.56$

$C_{uu} = $ max $($312,500–$220,000, 0) = $92,500$

$C_{ud} = $ max $($187,500–$220,000, 0) = 0$

$C_{dd} = $ max $($112,500–$220,000, 0) = 0$

Through Equation (7.2), we have

$C = 0.56^2($92,500)/1.03^2 = $27,343$

(2) If the lease is three years long, then

$C_{uuu} = $ max $($390,625–$220,000, 0) = $170,625$

$C_{uud} = $ max $($234,375–$220,000, 0) = $ \ 14,375$

$C_{udd} = $ max $($140,625–$220,000, 0) = 0$

$C_{ddd} = $ max $($ \ 84,375–$220,000, 0) = 0$

Through Equation (7.3), we have

$C = [0.56^3($170,625)+3(0.56)^2(0.44)($14,375)]/1.03^3 = $32,867$

We see that as the number of periods increases, the value of the Call option increases at a diminishing rate.

(3) A two-period project abandonment Put

Project abandonment is an important option for technology portfolio management. Such a Put option is valuable to a decision maker by allowing the right to sell a project at a fixed strike price in the future when the project is not performing as expected. The following example illustrates the estimation of the value of a two-period project abandonment Put option.

A technology project is currently valued at $1 million. At the end of each of the next two years, the value may go up or down by 10%. The technology developer has the option of selling the project to another company for $850,000 at the end of each year, which is the strike price. Risk-free interest rate is 5% per year. To estimate the value of this Put option, we use Equation (7.4) with the following data:

$u = 1.1, d = 0.9,$

$R = 1.05, q = (1.05–0.9)/(1.1–0.9) = 0.75$

$uS = 1.1$ million, $dS = 0.9 million,

$uuS = 1.21 million, $P_{uu} = $ max $(0.85–1.21, 0) = 0$

$udS = 0.99 million, $P_{ud} = $ max $(0.85–0.99, 0) = 0$

$ddS = 0.81 million, $P_{dd} = $ max $(0.85–0.81, 0) = $40,000$

$P = [(0.75)^2(0)+2(0.75)(0.25)(0)+(0.25)^2($40,000)]/(1.05)^2 = $2,368.$

Furthermore, the analysis indicates that the project should be abandoned when ddS occurs.

7.3.2.3. The Continuous Case—The Black-Scholes Formula

The formula

The derivation of the famous Black-Scholes formula for the value of the Call option of the continuous case is beyond the scope of this book. However, it is instructive to present the formula for the European style Call option:

$$C = e^{-DT} N(d_1) S - K e^{-rT} N(d_2)$$

where $d_1 = [\ln(S/K) + (r + 0.5\sigma^2)T]/(\sigma\sqrt{T})$

$d_2 = d_1 - \sigma\sqrt{T}$

D = dividend rate

r = risk-free interest rate

T = time remaining before expiration

$N(d)$ = probability that outcome of less than d will occur in a normal distribution with mean 0 and standard deviation 1

S = current market price of the underlying stock

K = exercise price of the option

σ = risk in the underlying stock in terms of volatility, or the standard deviation of the rate of the return on the stock

For the special case where D=0, i.e., no dividend paying, and T=0, i.e., at the time of contract expiration, we have $d_1 = d_2 = +\infty$, if S>K and $-\infty$, if S<K. Thus $N(d_1)$ = $N(d_2)$ =1, if S>K and 0 if S<K. As a result, we have C = S-K if S>K and 0 if S<K, which is consistent with the single period discrete binomial case.

Example

As an example, consider the lease option for a technology innovation currently priced at $200,000. Assume that D=0, r=0.03, T=1, K=$220,000, and σ=0.2482 (the standard deviation for a distribution with 56% probability for the return of 0.25 and 44% probability of for the return of –0.25, which is $[0.56(0.25–0.03)^2+0.44(–0.25–0.03)^2]^{0.5}$, with 0.03=0.56(0.25)+0.44(–0.25) being the mean return of the distribution). Then

$d_1 = \{\ln(200/220)+[0.03+0.5(0.2482)^2(1)]\}/0.2482 = -0.139$

$d_2 = d_1 - 0.2482(1) = -0.387$

From the standard normal distribution, we have $N(d_1)$ = 0.447 and $N(d_2)$ = 0.349. Thus,

$C = \$200,000(0.447)-\$220,000e^{-0.03}(0.349) = \$14,923$

This is comparable to the value of the Call option for the single period at $16,311 in Section 7.3.2.1.

7.3.3. Major advantages and disadvantages[14]

In investment decisions for technology development or application, there are many opportunities for real options. Traditional net present value approach to project valuation assumes a static future expected value for the project without active project management. As a result, it neglects the value of management flexibility and risk control. Applications of real options analysis correct this error.

Although similar in structure, classical decision tree analysis using a constant discount rate does not value the option properly. This is caused by the fact that the classical decision tree analysis uses probabilities for future price changes based on the judgments of the decision maker. These probabilities correspond to a specific risk-adjusted discount rate such that the present value of the expected price is the same as the current price. However, this risk-adjusted discount rate shifts because at the time of the contract expiration, the option value becomes dependent on whether the option is exercised or not. Real options analysis avoids this difficulty by using a replicating portfolio of the investment and a risk-free bond.

Use the specific example in Section 7.3.2.1, if the probabilities for u and d are judged to be 60% and 40% respectively, then the risk-adjusted discount rate r can be obtained from the following relationship:

$$
\begin{aligned}
\text{Current value} = 200,000 \ &= \ \text{Present value of future expected value} \\
&= \ [0.6(250,000)+0.4(150,000)]/(1+r) \\
r \ &= \ 0.2
\end{aligned}
$$

However, at the time of the option contract expiration, r is no longer the correct risk-adjusted discount rate, because the value of the option is either \$30,000 or 0 dependent on whether the option is exercised. The replicating portfolio approach of the real options analysis provides the correct present value of the option at \$16,311. Using this present value, we can use the above relationship again to back out the proper risk-adjusted discount rate ro for the decision-tree analysis:

$$
\begin{aligned}
16,311 \ &= [0.6(30,000)+0.4(0)]/(1+ro) \\
ro \ &= 0.1035
\end{aligned}
$$

On the other hand, if the risk-neutral probabilities $q=(1.03-0.75)/(1.25-0.75)=0.56$ and $1-q$ were used, then the risk-free interest rate, $r=0.03$, becomes the proper discount rate, as

$$16,311 = [0.56(30,000)+0.44(0)]/(1+0.03).$$

However, in spite of its introduction more than 25 years ago and extensive research in the academic community, applications of real options analysis is still very limited, particularly in technology management.

---

[14] Contents of this section are partly based on information from R. Fink, "Reality Check for Real Options," *CFO Magazine*, September 13, 2001.

One reason for the slow progress may be due to unfamiliarity and resistance from technical managers steeped in traditional engineering-economic analysis and classical decision tree analysis about financial analysis concepts.

Another reason appears to be the lack of meaningful and accurate data for determining either current or future prices of an emerging technology. Real options analysis relies heavily on the information of future market prices to estimate the value of the option. If there are no financial markets on which to base these prices, they are generally developed through judgments. If the judgments are faulty, then the valuation of the options will not be accurate and reliable. As a result, it is often difficult to determine the true value of flexibility of an option. These valuation and implementation difficulties also help explain why many companies use real options mainly on procurement and other aspects of its business where contracts are used.

Furthermore, real options work only if a company is truly prepared to take action to exercise the option by canceling a project if the results are not favorable after the initial investment. However, in reality, companies often have difficulty in implementing the results of the analysis because of social and political issues involved in project cancellation or abandonment. It is common knowledge that many companies continue to fund projects even when they no longer live up to expectations, if only because the managers involved have a vested interest in seeing them continue. Consequently, real options analysis must be supported by corporate discipline for the analysis to be effective. On the other hand, this difficulty also points out some inadequacy in properly modeling the cost involved in real options, which should include not only financial costs, but also social and political costs of implementing the results of the analysis.

Still another difficulty lies in the availability of sufficient capital funding for the development of options in the case of project expansion or new ventures. A downturn in the capital markets often make these real options unrealizable.

Given all these difficulties in accurate valuation and effective implementation, real options analysis at the present time is often more useful in conceptualizing projects and inducing organizational change than in evaluating and implementing them.

7.4. References

Decision Tree Analysis

Clemens, R. *Making Hard Decisions*, Duxbury 1997

Hammond, J., Keeney, R., and Raiffa, H., *Smart Choices*, Harvard Business School Press, 1998

Howard, R., *Dynamic Programming and Markov Process*, MIT Press, 1960

Kahneman, D. and Tversky, A., "Prospect Theory: An analysis of Decision Under Risk," *Econometrica*, 47, 276–287, 1979

Keeney, R., and Raiffa, H., *Decisions with Multiple Objectives*, Wiley, 1977

Luce, R., and Raiffa, H., *Games and Decisions*, Wiley, 1957

Raiffa, H. *The Art and Science of Negotiation*, Harvard University Press, 1982

Raiffa, H., *Decision Analysis*, Addison-Wesley 1968

Raiffa, H., and R. Schlaifer, R., *Applied Statistical Decision Theory*, Wiley, 1961

Stanford Research Institute, *Reading in Decision Analysis*, 1976

Zeckhauser, R., Keeney, R., and Sebenius, J., (Eds.), *Wise Choices*, Harvard Business School Press, 1996

Portfolio Diversification

Markowitz, H., *Portfolio Selection: Efficient Diversification of Investments*, Yale University Press, 1971

Markowitz, H., "Portfolio Selection," *Journal of Finance*, Vol. 7, No. 1, p. 77–91, 1952

Software: Software for portfolio diversification has been developed by Gopal Bhat and is available at the author's Web site www.starstrategygroup.com.

Real Options Theory

Black, F., and Scholes, M. "The Pricing of Options and Corporate Liabilities." *Journal of Political Economy,* Vol. 81, pp. 637–654, 1973

Copeland, T., and Antikarov, V., *Real Options: A Practitioner's Guide*, Thomson-Texere, 2003

Dixit, A.K. and Pindyck, N.S., *Investment Under Uncertainty*, Princeton University Press, 1994

Fink, R., "Reality Check for Real Options," *CFO Magazine*, September 13, 2001

Kester, W.C., "Today's Options for Tomorrow's Growth," *Harvard Business Review*, March-April, p.18, 1984

Luenburger, D.G., *Investment Science*, Oxford University Press, 1998

Merton, Robert C. "Theory of rational option pricing." *Bell Journal of Economics and Management Science*, Vol. 4, no. 1, pp. 141–183, 1973

Mun, J., *Real Options Analysis: Tools and Techniques for Valuing Strategic Investments and Decisions*, Wiley, 2002

Myers, S.C., "Determinants of Corporate Borrowing," *Journal of Financial Economics*, 5, p.147, 1977

Myers, S.C., "Finance Theory and Financial Strategy," *Midland Corporate Finance Journal*, p.5, 5, 1987

Pindyck, N.S., "Irreversible Investment, Capacity Choice and the Value of the Firm," *American Economic Review*, p.969, 79, 1988

Sharpe, W.F., *Investment*, Prentice Hall 1978

Trigeogis, L., *Real Options—Managerial Flexibility and Strategy in Resource Allocation*, MIT Press. 1994

7.5. Exercises

Problem 7.1

For the decision problem in Figure 7.6, compute the expected value of information for each of the following sets of conditional probabilities for the reliability of the consultant's forecast:

1. Perfect: $P(F|S)=P(U|N)=1$.
2. Perfectly wrong: $P(F|S)=P(U|N)=0$
3. Wishy-washy: $P(F|S)=P(U|N)=0.5$
4. Reliable: $P(F|S)=0.9$, $P(U|N)=0.8$
5. Unreliable: $P(F|S)=0.6$, $P(U|N)=0.7$

Problem 7.2

A manufacturer produces computers at the cost of $500 per unit and sells them for $1000 per unit. If a computer is produced but not sold, it has a salvage value of $200. Demand for computers from this manufacturer is estimated to be normally distributed with mean 100,000 units and standard deviation 1,000 units. What is the optimal number of units that the manufacturer should produce to maximize its expected net revenue?

Problem 7.3

Technology alternatives A and B have respective expected annual returns, μ's, and risks, expressed as the standard deviations, σ's of the investment returns, as follows:

$$\mu(A) = 30\% \qquad \sigma(A) = 20\%, \qquad \mu(B) = 10\% \qquad \sigma(B) = 5\%$$

An investor plans to invest a fraction p of her capital in A and the remaining fraction 1–p of her capital in B. The correlation coefficient between A and B is ρ. Compute the expected returns and risks for p = 0, 0.2, 0.4, 0.6, 0.8, and 1 and plot the returns and risks for these p values as a graph for $\rho = -1, -0.5, 0, 0.5$, and 1.

Problem 7.4

Develop as rigorous an argument as you can about why an iso-preference curve in the risk-return diagram are convex in that every point on the line connecting two points on an iso-preference curve will lie either on or above the curve.

Problem 7.5

The current market price of a product is $20. A year later, the price of the product is projected to have a 50–50 chance to be either $25 or $15. A company is considering:

1. spend $20 million at the beginning of the year to produce 1 million units of this product to be sold at the end of the year;
2. wait until the end of the year when the prevailing price of the product becomes known; if the price at that time is higher than the pre-set price of $22 million, then purchase 1 million units of the product from a manufacturer and sell it right away.

However, the manufacturer demands an option fee for the company to have the option of waiting until the end of the year. With the risk-free interest rate or discount rate for the company at 3% per year, what should be the upper bound for this fee?

Problem 7.6

Redo Problem 7.5 with each of the following sets of future prices:

1. $30 and $10
2. $40 and $0

Problem 7.7

Redo Problem 7.5 with the assumption that the option can be exercised at the end of each year for two years and find the value of the option at the beginning of the first year.

Problem 7.8

At the beginning of the year, a CEO plans to purchase a shipment of products from overseas. Depending on the conditions of the economy at the end of year, there are 80–20 chances of *instantly* generating revenues of uS=$125 million and dS=$80 million respectively when the products are delivered and sold at the end of the year. The seller currently asks $100 million for the shipment. The CEO would like to have the option of delaying the purchase until the latter part of the year when the economic condition is clear. If purchased at the end of the year, the agreed price for the shipment will be $110 million. The CEO can also invest in or borrow from a highly secure bond with 5% risk-free interest rate. What is the maximum amount that the CEO should pay for this lease?

Problem 7.9

A company invests $100 million at the beginning of the year in a new technology to be marketed at the end of the year. At that time, the technology has a 20% chance of producing $125 million and 80% chance of producing $80 million. However, another developer offers the company the option of selling the intellectual property of the technology at the end of the year for $90 million. The risk-free interest rate is 10%. What is the maximum amount the company should pay for this option?

8 Monitor Portfolio Progress and Changes

Once the technology portfolio has been chosen, the decision maker will need to track the progress of portfolio implementation. This chapter discusses two useful tools for tracking such progress: the Project Evaluation and Review Technique and a Project Portfolio Monitoring System.

8.1. Project Evaluation and Review Technique

Project Evaluation and Review Technique (PERT) and an allied tool, the Critical Path Method (CPM), were developed in the 1950s in support of project management for large defense programs. These tools were used to develop a network diagram delineating the flow of project activities from start to finish, and to provide an algorithm for identifying and modifying the set of activities that controls the overall project completion time. Over the years, they have been widely applied to many other technology development and construction programs, with well-established sophisticated computer software. In this section, we will focus on the underlying general principles of these tools. Furthermore, the artificial distinction between PERT and CPM is that PERT generally emphasizes the probability distribution of the project completion time, while CPM usually emphasizes the modification of the expected project completion time. However for simplicity's sake, we will combine both features under our discussion of PERT.

PERT can be used to manage a technology development process through the following steps.

8.1.1. Collect data on project activities

The first step of the process is to collect project activity data from the project planner, the project manager, and/or experienced contractors. The basic data include:

- the sequencing of all project activities
- the probability distributions of the completion times of individual activities, which are often observed as modified beta distributions that can be characterized by the most optimistic, likely, and pessimistic completion times
- the normal costs of completing individual activities

Other relevant data useful for project management include:

- the additional expected times that can be shortened or lengthened for individual activities
- the additional costs for shortening or lengthening individual activities

As a specific example, the basic data for the simplified activities for a typical technology development are given in Table 8.1 below.

Table 8.1. Basic Data for a Generic Technology Development

Activity	Immediate Predecessor	Activity Completion Time (months)		
		Optimistic a	Likely m	Pessimistic b
A. Proof of physical feasibility	—	1	2	3
B. Proof of engineering feasibility	A	1	3	5
C. Scalability and market assessment	A	2	4	6
D. Preliminary system design	B,C	2	4	6
E. Proof of concept and pilot study	D	3	6	9
F. Prototype design and development	E	6	9	18
G. Production facility construction	F	12	15	24
H. Marketing and sales planning	C,F,G	4	6	8
I. Production rollout	G	9	12	21
J. Marketing and sales campaign	H	4	6	8

8.1.2. Estimate the expected values and variances of individual activity completion times

Estimates of activity completion time have been empirically observed to follow a beta probability distribution. Based on the characteristics of the beta distribution, the expected value and variance of the activity completion time can be estimated as follows:

Expected activity completion time = $E(t) = (a+4m+b)/6$

Variance of activity completion time = $V(t) = [(b-a)/6]^2$

These estimated expected values and variances of the activity completion times in the example above are summarized in Table 8.2.

8.1.3. Construct the PERT diagram

With information from Table 8.2, we can construct the PERT diagram as follows:

Table 8.2. Expected Values and Variances of Activity Completion Times

		Completion Time t	
Activity	Immediate Predecessor	Expected Value E(t)	Variance V(t)
A. Proof of physical feasibility	—	2	1/9
B. Proof of engineering feasibility	A	3	4/9
C. Scalability and market assessment	A	4	4/9
D. Preliminary system design	B,C	4	4/9
E. Proof of concept and pilot study	D	6	1
F. Prototype design and development	E	10	4
G. Production facility construction	F	16	4
H. Marketing and sales planning	G	6	1/9
I. Production rollout	G	13	4
J. Marketing and sales campaign	H	6	4/9

- For activities with no immediate predecessor, they will be preceded by the START node.
- All activities with no following activities will go to the FINISH node.
- Each activity is represented by a three-row rectangular box with the first row showing the activity symbol and expected completion time.
- The second row of each activity box lists ES, the earliest starting time for the activity, which is the *maximum* of EF's of all preceding activities because the activity cannot start until all preceding activities have finished; and the earliest finishing time, EF, which is ES+E(t)
- The earliest starting time ES of activities immediately following the START node is 0
- Let EFP be the maximum of EF's of activities directly followed by the FINISH node, then the third row of each of these activity boxes will list the latest starting time LS=EFP-E(t) and the latest finishing time LF=EFP.
- Going backwards, the LF of each activity box will be the *minimum* of LS of all following activities, because this activity must finish when one of the following activities needs to start; and LS=LF-E(t)

Following the above algorithm, the PERT diagram for the example is given in Figure 8.1.

8.1.4. Identify the critical activities, critical path, and expected project completion time

The critical activities are those activities for which ES=LS or equivalently EF=LF. These activities form the Critical Path for the project and control the expected project completion time. It should be noted that it is possible to have two or more critical paths with equal expected project completion time.

In the above example, the critical path is the chain formed by the critical activities A, C, D, E, F, G, and I as indicated by the boxes with boldfaced letters. The expected project completion time, E(T), is then 55 months.

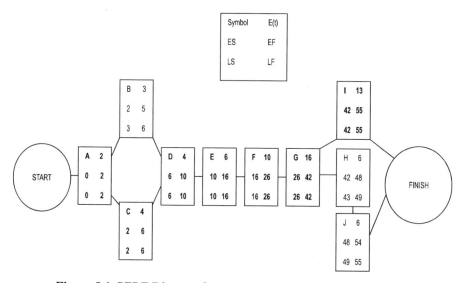

Figure 8.1. PERT Diagram for a Generic Technology Development

8.1.5. Estimate variance of project completion time and probability of project completion by a given time

For a specific critical path, the variance of the project completion time, V(T), can be estimated as:

$$V(T) = \Sigma_i \Sigma_j \, \rho_{ij} \, \sigma_i(t) \, \sigma_j(t)$$

where $\sigma_i = \sqrt{V_i(t)}$ = standard deviation of activity i

ρ_{ij} = correlation coefficient between activities i and j

For the generic technology development example with statistically independent activities; i.e., $\rho_{ij} = 1$ for i=j and 0 otherwise,

$$V(T) = (1/9)+(4/9)+(4/9)+1+4+4+4=14 \text{ month}^2$$

If multiple critical paths are present, theoretically the expected project completion time should be the expected value of the *maximum* of the project completion times of all critical paths, which is generally not the same as the common expected project completion time of individual paths. The variance of the project completion time has an even more complex relationship with the variances of the completion times of individual paths. For simplicity, we will approximate the expected project completion time with multiple critical paths by the common expected project completion time of individual paths, and approximate the variance of the project completion time by the largest variance among these critical paths.

Now if the activities are statistically independent and the number is 30 or more, then by the Central Limit Theorem in statistics, the project completion time will be approximately normal distributed with mean $E(T)$ and variance $V(T)$. Thus, we can estimate the probability of project completion by a given time t through the following formula:

$$P\{T<t\} = P\{Z<[t-E(T)]/\sigma(T)\}$$

where Z is the random variable with a standard normal distribution

$$\sigma(T) = \sqrt{V(T)} = \text{the standard deviation of the project completion time.}$$

For the above example, if $t = 50$ months, then

$$P\{T< 50\} = P\{Z<(50-55)/\sqrt{14}\} = P\{Z<-1.34\} = 0.09.$$

8.1.6. Develop project budgeting

The PERT diagram can be used to develop a budgeting process for the project. Again using the generic technology development as the example, Table 8.3 lists the normal costs of individual activities obtained from an experienced contractor as well as the normal time $E(t)$ from Table 8.2.

Using the earliest and the latest starting times from Figure 8.1, the cumulative budget expenditure requirements over time are summarized in Table 8.4.

Table 8.3. Normal Costs of Individual Activities

Activity	Normal Time E(t) (months)	Normal Cost C ($ million)	Cost per Month C/E(t) ($ million)
A	2	0.5	0.25
B	3	0.9	0.30
C	4	1.0	0.25
D	4	2.0	0.50
E	6	3.6	0.60
F	10	4.0	0.40
G	16	8.0	0.50
H	6	0.6	0.10
I	13	7.8	0.60
J	6	2.4	0.40
Total		30.8	

Table 8.4 indicates that there is room for delaying expenditures in weeks 3–5 and 43–54, which can be helpful in cases of tight cash flows.

Table 8.4. Cumulative Budget Expenditure Requirements

Activity	1	2	3	4	5	6	10	16	26	42	43	48	49	54	55
								Week							
A	0.25	0.5	0.5	0.5	0.5	0.5	0.5	0.5	0.5	0.5	0.5	0.5	0.5	0.5	0.5
B-ES			0.3	0.6	0.9	0.9	0.9	0.9	0.9	0.9	0.9	0.9	0.9	0.9	0.9
B-LS				0.3	0.6	0.9	0.9	0.9	0.9	0.9	0.9	0.9	0.9	0.9	0.9
C			0.25	0.5	0.75	1.0	1.0	1.0	1.0	1.0	1.0	1.0	1.0	1.0	1.0
D							2.0	2.0	2.0	2.0	2.0	2.0	2.0	2.0	2.0
E								3.6	3.6	3.6	3.6	3.6	3.6	3.6	3.6
F									4.0	4.0	4.0	4.0	4.0	4.0	4.0
G										8.0	8.0	8.0	8.0	8.0	8.0
H-ES											0.1	0.6	0.6	0.6	0.6
H-LS												0.5	0.6	0.6	0.6
I											0.6	3.6	4.2	7.2	7.8
J-ES													0.4	2.4	2.4
J-LS														2.0	2.4
Total-ES	0.25	0.5	1.05	1.6	2.15	2.4	4.4	8.1	12.0	20.0	20.7	24.2	25.2	30.2	30.8
Total-LS	0.25	0.5	0.75	1.3	1.85	2.4	4.4	8.1	12.0	20.0	20.6	24.1	24.8	29.8	30.8

8.1.7. Analyze project completion time reduction and extension[15]

An early completion of the project may provide competitive advantages, such as early market entry and profit generation. In this case, the decision maker may wish to reduce the project completion time by reducing the activity completion times. However, to reduce the completion time of an activity will require additional costs. Furthermore, because it generally becomes increasingly difficult to compress the completion time of an activity, the amount of additional cost required would increase with the time to be reduced and eventually reach infinity when it becomes physically impossible to reduce the activity completion time any further.

On the other hand, it can sometimes be advantageous to extend the project completion time by relaxing the activity completion times so that the required completion costs for the activities may be reduced. In general, the incremental reduction in cost for an activity will gradually diminish and eventually reach zero when there can be no further reduction in activity completion cost no matter how much time extension is allowed for the activity.

As a result, the cost of completing an activity is typically a convex function of the activity completion time as shown in Figure 8.2.

---

[15] Concepts for project completion time extension are based on original work by the author.

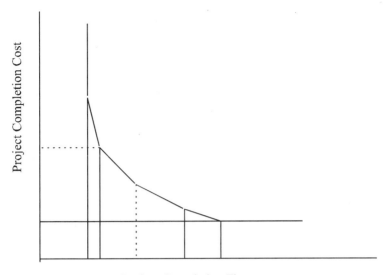

Project Completion Time

Figure 8.2. Typical Project Completion Cost as a
Function of Project Completion Time

With data on allowable changes in completion time and the associated cost for each activity, we can analyze the tradeoff between changes in project completion time and cost. Specifically, to reduce project completion time by one unit at an additional cost, we must select the critical activity with the least cost increase per unit of time reduced, since reducing non-critical activities does not affect project completion time.

On the other hand, to reduce project completion costs with potential increases in project completion time, we should first select the non-critical activity with the most decrease in cost per unit of time lengthened, as lengthening a non-critical activity can reduce project completion costs without increasing project completion time. If no non-critical activity is available, we will apply the this criterion to all activities.

It is important to note that each reduction of project completion time is likely to turn a non-critical activity into a critical activity and result in additional critical paths. Furthermore, to reduce the completion time with multiple critical paths, we must either reduce the completion time of an activity common to all paths or that of one activity from each path; otherwise, the project completion time will not be affected. Given these characteristics, we should reduce the completion time *one unit at a time* so that we will not miss any new critical path and we need to carefully search among all critical paths the combination of activities that will provide the least cost increase for the reduction.

For the generic technology development example, the allowable changes in activity completion time and cost obtained from an experienced contractor are summarized in Table 8.5.

To reduce the project completion time by one month, i.e., from 55 months to 54 months, we will choose to reduce C, the critical activity with the least cost increase per month reduced, by one month for an incremental cost of $0.25 million.

Table 8.5. Allowable Changes in Activity Completion Time and Cost

Activity	Minimum Time (months)	Maximum Cost ($ million)	Cost Increase per Month Reduced ($ million)	Normal Time (months)	Normal Cost ($ million)	Maximum Time (months)	Minimum Cost ($ million)	Cost Decrease per Month Extended ($ million)
A	1	1.1	0.6	2	0.5	3	0.3	0.2
B	1	1.5	0.3	3	0.9	5	0.6	0.15
C	2	1.5	0.25	4	1.0	6	0.7	0.15
D	2	3.0	0.5	4	2.0	6	1.5	0.25
E	3	6.0	0.8	6	3.6	9	2.4	0.4
F	6	6.4	0.6	10	4.0	18	3.0	0.125
G	12	10.6	0.65	16	8.0	24	6.0	0.25
H	4	1.0	0.2	6	0.6	8	0.4	0.1
I	9	10.4	0.65	13	7.8	21	6.0	0.225
J	4	3.5	0.55	6	2.4	8	2.0	0.2

After this reduction of project completion time, the non-critical activity B now becomes critical, and there are two critical paths: A-B-D-E-F-G-I and A-C-D-E-F-G-I. Thus, at this stage, reducing the completion time of critical activity C alone will no longer reduce the project completion time. To reduce the project completion time by one more month, we will then need to reduce either the completion times of both critical activities B and C or that of one of the common critical activities, A, D, E, F, G, or I. We will now choose to reduce the common critical activity D by one month, as it offers the least cost increase of $0.5 million for reducing the second month.

Following this process, the additional reductions in project completion time with the lowest cost, together with the first two reductions, are summarized in Table 8.6.

On the other hand, we can extend the completion times of non-critical activities J and B by one month each without increasing the project completion time, while reducing the total project cost by $0.35 million. However, at this stage, all activities become critical. To extend further, we will choose to extend activities I and J, which produce the most cost decrease of $0.425 million in extending the project completion time by one month.

Following this process, the additional increases in project completion time with the most cost reduction, together with the first three extensions, are summarized in Table 8.7.

Table 8.6. Reductions in Project Completion Time with the Lowest Cost

Activities with Completion Time Reduced by One Month	Project Completion Time (months)	Incremental Cost ($ million)	Project Completion Cost ($ million)
C	54	0.25	31.05
D	53	0.5	31.55
D	52	0.5	32.05
B,C	51	0.55	32.60
A	50	0.6	33.20
F	49	0.6	33.80
F	48	0.6	34.40
F	47	0.6	35.00
F	46	0.6	35.60
G	45	0.65	36.25
G	44	0.65	36.90
G	43	0.65	37.55
G	42	0.65	38.20
I	41	0.65	38.85
E	40	0.8	39.65
E	39	0.8	40.45
E	38	0.8	41.25
H,I	37	0.85	42.10
H,I	36	0.85	42.95
I,J	35	1.2	44.15
	Total	13.35	

The changes in project completion cost as a function of project completion time resulting from both reduction and extension are summarily shown in Figure 8.3.

8.1.8. Use linear programming for project completion time reduction

Because the PERT diagram can be viewed as a network of time flows, project completion time reduction can also be conducted by applying the linear programming technique as follows:

For Activity i:

Let X_i be the earliest finishing time (EF), with $X_0=0$ being the EF of the START node, and X_n being the EF of the FINISH node, or the project completion time, and let T be the target to which X_n is to be reduced.

Let Y_i be the amount of the completion time to be reduced with $Y_n=0$ for the FINISH node, and let C_i be the incremental cost per unit of completion time reduced.

Let N_i and m_i be respectively the normal and minimum activity completion times.

Table 8.7. Extensions in Project Completion Time with the Cost Reduction

Activities with Completion Time Increased by One Month	Project Completion Time (months)	Incremental Cost Reduction ($ million)	Project Completion Cost ($ million)
B,J	54	0.35	30.45
I,J	55	0.425	30.025
E	56	0.4	29.625
E	57	0.4	29.225
E	58	0.4	28.825
H,I	59	0.325	28.5
H,I	60	0.325	28.175
B,C	61	0.3	27.875
D	62	0.25	27.625
D	63	0.25	27.375
G	64	0.25	27.125
G	65	0.25	26.875
G	66	0.25	26.625
G	67	0.25	26.375
G	68	0.25	26.125
G	69	0.25	25.875
G	70	0.25	25.625
G	71	0.25	25.375
I	72	0.225	25.15
I	73	0.225	24.925
I	74	0.225	24.7
I	75	0.225	24.475
I	76	0.225	24.25
A	77	0.2	24.05
C	78	0.15	23.9
F	79	0.125	23.775
F	80	0.125	23.65
F	81	0.125	23.525
F	82	0.125	23.4
F	83	0.125	23.275
F	84	0.125	23.150
F	85	0.125	23.025
F	86	0.125	22.9

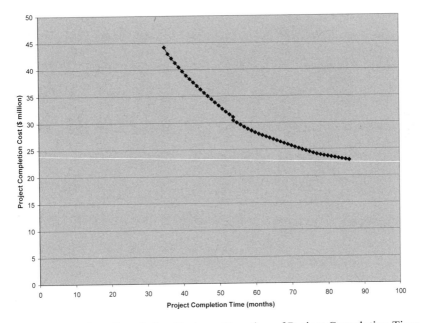

Figure 8.3. Project Completion Cost as a Function of Project Completion Time

Then the linear program for project completion time reduction is:

Minimize	$\Sigma C_i Y_i$	
Subject to	$0 \leq Y_i \leq N_i - m_i$	for $i=1,2,\ldots,n-1$
	$X_i \geq X_{pi} + (N_i - Y_i) \geq 0$	for $i=1,2,\ldots,n$
	$X_n \leq T$	

where X_{pi} is the earliest finishing time of a predecessor of Activity i.

For the generic technology development example, let T=35, and use the alphabets rather than the numerals as the indexes for Y and the data from Table 8.5, then the linear program is:

Minimize $C=0.6YA+0.3YB+0.25YC+0.5YD+0.8YE+0.6YF+0.65YG+$
            $0.2YH+0.65YI+0.55YJ$

Subject to $0 \leq YA \leq 2-1=1$
            $0 \leq YB \leq 3-1=2$
            $0 \leq YC \leq 4-2=2$
            $0 \leq YD \leq 4-2=2$
            $0 \leq YE \leq 6-3=3$
            $0 \leq YF \leq 10-6=4$
            $0 \leq YG \leq 16-12=4$
            $0 \leq YH \leq 6-4=2$
            $0 \leq YI \leq 13-9=4$
            $0 \leq YJ \leq 6-4=2$
            $XA \geq 0+(2-YA) \geq 0$

$$XB \geq XA+(3-YB) \geq 0$$
$$XC \geq XA+(4-YC) \geq 0$$
$$XD \geq XB+(4-YD) \geq 0$$
$$XD \geq XC+(4-YD) \geq 0$$
$$XE \geq XD+(6-YE) \geq 0$$
$$XF \geq XE+(10-YF) \geq 0$$
$$XG \geq XF+(16-YG) \geq 0$$
$$XH \geq XG+(6-YH) \geq 0$$
$$XI \geq XG+(13-YI) \geq 0$$
$$XJ \geq XH+(6-YJ) \geq 0$$
$$Xn \geq XI$$
$$Xn \geq XJ$$
$$Xn \leq 35$$

Solving this linear programming yields C=44.15.

8.1.9. Major advantages, limitations, and applicability

Since their initial development in 1950s, PERT and CPM have been widely applied in all major technology development and facility construction project. The concepts are simple, straightforward, and easy to understand. Furthermore, sophisticated and easy-to-use computer software programs are commercially available.

However, this tool requires detailed understanding of the precedence relationships of the activities, which may change over time. Moreover, careful considerations need to be made in the selection of a critical activity in the shortening process of the project completion time, as the activity selected can have complex repercussions on future evolution of the critical path and actual project operation.

8.2. Project Portfolio Monitoring System: SRI Early Alert System[16]

For a technology portfolio with many projects, it will be important for the decision maker to monitor the progress of these projects in order to make timely modifications if the need should arise. One such portfolio monitoring system is the SRI Early Alert System (SEAS), first developed by Thomas Boyce at SRI International in 1998. SEAS combines the use of expert experience, structured argumentation, and collective reasoning to monitor changes and extract warning signs from these observable changes. Over the years, it has been renamed Structured Evidential Argumentation System and has evolved at the Artificial Intelligence Center of SRI into a much more sophisticated and complex software that has been applied to government and commercial operations far beyond the original project portfolio monitoring and management purposes.

---

[16] The description of SEAS in this section is based on publicly available information from SRI International. The illustrative example on project portfolio monitoring was developed by the author.

In the following section, we will discuss the basic concepts of SEAS and provide a simple illustrative example in its application to project portfolio monitoring and management.

8.2.1. Underlying concepts of SEAS

Conceptually, SEAS combines a project status reporting methodology based on expert input with a computer software using artificial intelligence techniques to automatically assess the need for management intervention in a portfolio of projects. The output from the system is analogous to that of a warning light on an automobile dashboard. It signals an incipient problem early enough to allow for corrective action prior to costly damages.

More specifically, SEAS uses a standard set of questions to periodically assess key aspects of both internal and external situations, as a means of comparing and contrasting their impact on the entire project portfolio. The system supports a hierarchy of questions and answers with a hierarchy of screen images. At the top of this hierarchy, the screen displays the key questions to be asked and answered. The semi-quantitative answers to these questions are portrayed alongside each question as a sequence of lights ranging from green to yellow to red. A green light corresponds to a favorable response to a question, while yellow corresponds to a less favorable response and red to an unfavorable response. If one light is on, that is the answer; if multiple lights are on, it indicates that the answer is one of those that are lit, but there is not enough information at present to give a definitive answer; by convention, if all answers remain possible given current information, no lights are lit. The advantage of this display is that the answers to all of the key questions can be quickly scanned to understand the overall characteristics of the assessment and reveal which questions require additional attention.

By selecting any of these questions/answers for a more detailed examination, a screen image appears that repeats the selected question/answer at the top and displays the supporting questions/answers for this question, just below it. It is the answers to these supporting questions in the hierarchy that are driving the answer to the selected focus question. Again, this display allows the user to quickly see how the answer to the focus question is derived from the answers to its supporting questions. Selecting any of these supporting questions/answers will reveal a similar screen image for its supporting questions, or will reveal a screen where that question is directly answered. In the latter case, the screen repeats the question and provides a list of possible answers. By clicking on one or more of these, the user enters an answer for the question. Other fields on the screen record the date/time, the name of the user making the entry, and any reference material on which the answer is based. Collectively these entries provide a summary of the information that supports the answer to the question.

When the user answers previously unanswered questions, or changes the answers to others, the answers to questions higher in the hierarchy are automatically updated to reflect the new information. Evidential reasoning techniques of artificial intelligence together with human judgments are then used together to interpret the implications of the new information and decide on a course of management interventions and actions.

The graphical display of the general structure of SEAS has been extracted from the SRI International Web site and shown in Figure 8.4 below.

Structured Argumentation

▼ Structured arguments & templates record analytic products and methods

▼ Templates structure and guide analytic thinking encouraging higher fidelity reasoning

▼ Cascaded templates support deeper reasoning where the analyst determines that it is desirable

▼ Structured arguments are easily communicated, explained, and compared

▼ Structured arguments record lines of reasoning allowing users to drill down to supporting documentary evidence, its relevance, the answers it induces, and the rationale for those answers

▼ Graphical depictions of structured arguments speed comprehension and comparison

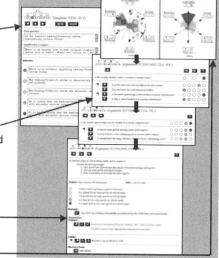

Collective Reasoning

▼ Web server architecture
  ▪ Simultaneous access
  ▪ Across all browser clients
  ▪ Little or no systems integration

▼ Published arguments/templates are guaranteed to be both stable and persistent
  ▪ The audience has simultaneous read access
  ▪ The co-authors have simultaneous write access, while unpublished, & read access thereafter

▼ Memos attached to arguments/templates provide a means for users to communicate asynchronously
  ▪ critiques, instructions, to-do, ...

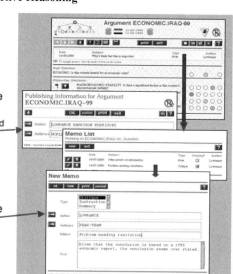

(continued)

Figure 8.4. The Structure of SEAS

Corporate Memory

▼ A knowledge base of analytic products (arguments) & analytic methods (templates), indexed by the situations to which they apply

▼ Arguments are the opinions that drove decisions; Templates capture best practice

▼ Expanded by analysts as a by-product during productive use of the tool rather than being required before the tool becomes useful

▼ Queries against corporate memory retrieve arguments/templates that were applied to similar situations

▼ Summaries of retrieved arguments/templates allow one to quickly understand the thinking of the past and how it might apply to the present

▼ Guaranteed access/privacy, persistence, and information assurance

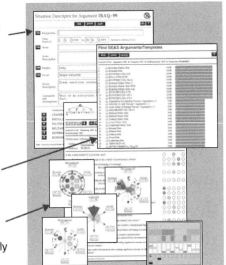

Figure 8.4 *(concluded)*. The Structure of SEAS

8.2.2. A Simple Illustrative Example

In the following simple example, we will illustrate the general principle of SEAS in monitoring the progress of a portfolio of projects by outlining the major elements of such a system:

(1) Project questionnaire[17]

A short multiple-choice questionnaire is developed to gauge the individual project progress in a class of similar projects based on various observable evidence of the project performance over time. This questionnaire is designed using both expert project management experience and evidential reasoning process. The evidence to be collected by the questionnaire includes:

- Resource utilization
  - Human resources
  - Financial resources

---

[17] The questionnaire can also be developed through the use of computer-assisted brainstorming software such as Angler™ to collect diverse inputs and systematically combine them into major categories of structured questions that are supported by corporate experience, expert opinions, and logical reasoning.

- Operational adequacy
  - Technical developments
  - Operational interactions with stakeholders inside and outside the organization

A sample of typical questions included in the questionnaire is given below.

Dimension 1: Resource Utilization
1.1 Staff availability on schedule?
1.2 Staff expertise sufficient?
1.3 Support effort sufficient?
1.4 Equipment and services sufficient?
1.5 Budget expenditure on schedule?
1.6 Contingency funds sufficient?

Dimension 2: Operational Adequacy
2.1 Technical developments on schedule?
2.2 Significant technical innovations?
2.3 Prototype operational?
2.4 Technical problems?
2.5 Production cost competitive?
2.6 Product application as planned?
2.7 Benefits as planned?
2.8 Contact with marketing group inside the organization?
2.9 Relation with external markets?
2.10 Contact and relationship with external sponsors?

(2) Project baseline plan

For each project in the same class, based on the explicit stated objectives and implementation plan of the project, a baseline plan is established upon project initiation. This baseline plan will include the milestones and expectations for the resource utilization and operational adequacy of the project.

(3) Data collection

Project data are then collected on a periodic basis with each project manager filling out the short questionnaire of multiple choice questions with answers on a scale of 1 to 10. This questionnaire further requires the project leader to make both qualitative and quantitative estimates of the likelihood of project success relative to the baseline plan.

(4) Data analysis

Artificial intelligence techniques are then used to automatically analyze the data collected from all projects in the same class and provide a report of potential warning signs. A simple technique would be displaying a color indicator for the performance of a project in the report by counting the number of problematic responses in the questionnaire. When the number is below a predetermined first threshold level, the color will remain green. When it exceeds the first threshold level, then the color turns yellow and attention by the project manager is warranted. Finally, if the number exceeds a second threshold level, then the color will turn red, and intervention by upper management would be required.

(5) Management actions

The data analysis is designed to prompt management actions. In the case of color signals, the project manager's responsibility is to take action to turn the color from yellow to green. A red signal would indicate the problem in the project development is sufficiently serious that upper management intervention would be required.

8.2.3. Major advantages, limitations, and applicability

The basic concepts and principles of the project portfolio monitoring system are intuitively appealing and have been practically applied not only to technology portfolio management but also to many other fields such as intelligence analysis and fraud detection. Furthermore, such a system provides a continuous record that encourages honest assessment and reporting on the part of individual project managers, while at the same time provides a clear and concise summary report that pinpoints problematic aspects of projects within the technology portfolio to the senior management. A primary benefit of such a system is that projects can be replanned at a much earlier stage, thereby making better use of resources and improving results.

The major challenges in developing and applying such a system are (1) in the design of the questionnaire that would be both effective in extracting the early warning signs and easy for project managers to complete meaningfully in a short time, and (2) in the reliability and validity of the evidential reasoning technique for using new information to correctly answer the specific questions.

At SRI International, SEAS has been continually refined, expanded, and used as an aid to institute-wide project management, monitoring of defense intelligence, and detection of tax fraud.

8.3. References

Project Evaluation and Review Technique

Stires, D., *Modern Management Methods PERT and CPM: Program evaluation review technique and critical path method,* Materials Management Institute, 1963.

Software: Software for PERT project crashing using linear programming has been developed by Gopal Bhat and is available at the author's Web site, www.starstrategygroup.com.

Project Portfolio Monitoring System

Daichendt, G. and Johnson, B., *I-Operations: The Impact of the Internet on Operating Models,* The Institute Press, 2000

8.4. Exercises

Problem 8.1

A technology development project has the following activities and related data:

		Time (weeks)			Crash	
Activity	Predecessor	Optimistic	Most Probable	Pessimistic	Cost/week ($1000)	Crashable Weeks
A	-	4	6	14	1	1
B	-	4	6	20	3	1
C	A,B	2	4	6	7	2
D	C	8	10	12	4	2
E	C	1	4	7	2	1
F	E	4	5	6	3	1
G	C	4	6	14	1	2
H	D,F,G	4	6	8	7	1

(a) Compute the mean and variance of each activity.
(b) Develop the PERT diagram and identify the critical path.
(c) What are the mean and standard deviation of the project completion time?
(d) Assuming a normal distribution for the project completion time, what is the probability that the project will take more than 0.5 standard deviation above the mean time to complete?
(e) Find time T so that there is a 95% probability for the project to be completed by T.
(f) Compute the incremental costs of reducing the mean project completion time for 4 weeks by crashing it one week at a time.

Problem 8.2

Develop the linear programming formulation of Problem 8.1(f) and use an available LP code to find the incremental costs of reducing the mean project completion time by 4 weeks.

Problem 8.3

Design a set of questionnaires for managing projects in your business or personal life, such as study, work, health pursuits, family activities, recreational activities, etc. Use it to develop a project portfolio monitoring system and conduct a trial application.

9 Modify Portoflio to Re-Integrate with Organizational Strategy

The technology portfolio planning and management process starts with the values as expounded in the organizational strategy, which is followed by the development of a technology portfolio in support of the strategy. Now with changes in the business environment as well as in the technology developments, not only the technology alternatives may be outdated, but also the organizational strategy may have been revised. Therefore, the technology portfolio will need to be modified to reconfirm its optimality and to reintegrate with the revised organizational strategy. This chapter will present two qualitative tools for portfolio modification and reintegration.

9.1. Factor Analysis

Factor analysis is a simple, intuitive, and widely used tool for examining the strengths and weaknesses of individual technologies in a portfolio. The analysis is carried out through the following steps:

9.1.1. Select and define factors

The first step of the analysis is to select and clearly define a set of factors that represent the overall values of a technology in the portfolio to the decision maker. To avoid double counting, these factors should be as uncorrelated with one another as practical. Typical factors include:

- *Strategic Importance*—Importance of the technology as a sustained competitive advantage for the organization
- *Commercial Value*—Size of the financial impact to the organization if the technology is commercially successful
- *Commercial Timing*—Time at which the market will adopt or purchase the technology at an acceptable business volume
- *Risks*—Likelihood that the technology will fail to accomplish its technical objectives, and that, if technically successful, it will fail commercially
- *Current Position*—Strength and ability of the company versus competitors in developing the technology today
- *Technology Availability*—Availability of technology from any source for commercialization

9.1.2. Develop detailed specifications and rating measures of the factors

In addition to qualitative definitions of these factors, we can develop more detailed specifications and ordinal or numerical rating measures of the factors based on these detailed specifications. Using the examples of the factors discussed above, we have the following more detailed specifications:

- *Strategic Importance*—Degree of impact based on market share, product differentiation, cost efficiency, and market entry speed
- *Commercial Value*—Net present value, return on investment, revenue from increased sales, and other financial measures
- *Commercial Timing*—Calendar time in years with estimated probability
- *Risks*—Probabilities of technical and commercial failures based on internal capability and resource availability and external market size, position, and future uncertainty
- *Current Position*—Degree of strength based on past experience, existing patents, and current capability of the company versus competitors
- *Technology Availability*—Number of sources and their willingness to license

The bases for the typical ordinal rating measures of these factors are given in Table 9.1.

Table 9.1. Bases for Typical Ordinal Rating Measures of the Factor Examples

Rating Measure	Importance (market impact)	Value (net present value)	Timing (years)	Risks (failure probability)	Position	Availability
			Factor			
High	Major, broad	>$500 million	0–2	<10%	World leader	Readily
Medium	Significant in some key segments	$50–500 million	3–7	10–30%	Major follower	Limited
Low	Minor or isolated	<$50 million	>7	>30%	Not competitive	None

If numerical rating measures are desired to achieve greater precision, the Analytic Hierarchy Process and other tools presented in Chapter 2 can also be applied.

It is important to note that these factors should be evaluated within the same scenario of future business environment with respect to market condition, business competition, and organizational strategy.

9.1.3. Display assessment of the factors of a technology

Using these rating measures, the strengths and weaknesses of various factors of a technology in the portfolio can be displayed graphically in a spoke-type diagram as shown in Figure 9.1.

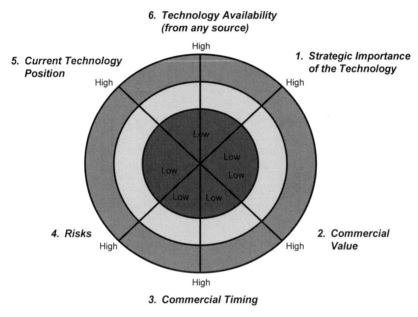

Figure 9.1. Graphical Display of Factor Analysis

The outer rim of the circle in Figure 9.1 represents the most desirable ratings of the factors of the technology. A technology with factors that have ratings falling below the outer rim would indicate that either those factors should be improved or a complementary technology could be combined with this technology to form a portfolio with more desirable combined ratings for all factors.

As an example, the factor analysis for a technology portfolio with high strategic importance, high commercial value, moderate market timing, unfavorable technical and financial risks, moderate competitive position, and moderate availability is shown in Figure 9.2. The diagram indicates that the technology portfolio is important and valuable, but there need to be ways to mitigate the high risks. Furthermore, efforts should be made to improve marketing time, competitive position, and availability.

9.1.4. Conduct an integrated factor analysis

The factor analyses of individual technologies in a portfolio under different scenarios can be integrated to reveal the *robustness* of the portfolio of technologies across different scenarios and the *strength* of the portfolio within each scenario. Table 9.2 provides an illustrative example of a simple, qualitative integrated analysis of the strategic importance of a portfolio of technologies under different scenarios.

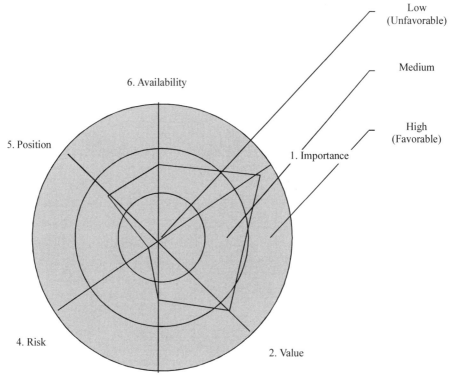

Figure 9.2. An Example of the Factor Analysis of a Technology Portfolio

Table 9.2. An Example of Integrated Qualitative Ratings of Technologies in a Portfolio for a Factor Under Different Scenarios

Factor 1—Strategic Importance of the Technology Portfolio				
	Scenario			
Technology in Portfolio	A	B	C	Overall Rating
1	MH	MM	MH	MH
2	HL	LH	MH	MM
3	LM	LM	LM	LM
4	MH	ML	LH	MM
Etc				
Portfolio Rating	MM	LH	ML	MM

In addition to the simple qualitative approach, there are many other more sophisticated ways to develop these ratings in the integrated factor analysis.

For example, in terms of Importance, Value, and Position, the portfolio rating can be estimated as a weighted average of the ratings of individual technologies. The specific weights can be developed judgmentally or through analytical methods, such as the Analytic Hierarchy Process presented in Chapter 2.

On the other hand, in terms of Timing, Risk, and Availability, the portfolio ratings can be estimated from an integration of the rating probability distributions of individual technologies. Again, the specific integration can be carried out by judgments or through more formal analysis like the portfolio risk assessment presented in Chapter 7.

Again, the degree of analytical sophistication should strike a proper balance between the value added and the resources required.

9.2. Strategy Map

Strategy map is another simple and useful tool for examining the *interactions and balances* between two factors for each technology alternative in a given scenario. These interactions and balances can then be used to provide strategic directions for technology portfolio development and modification.

9.2.1. Develop strategy maps and identify action areas[18]

For each pair of factors, we can construct a strategy map, which is essentially a two-dimensional matrix with ordinal rating levels for each dimension. The different combinations of rating levels of the two dimensions or factors can be used to not only gain insights about the implications of the combinations but also identify the areas for action if technologies fall in areas represented by the combinations. The following annotated figures provide a number of illustrative examples.

---

[18] The author is indebted to Thomas Boyce for developing the annotated strategy maps in this chapter.

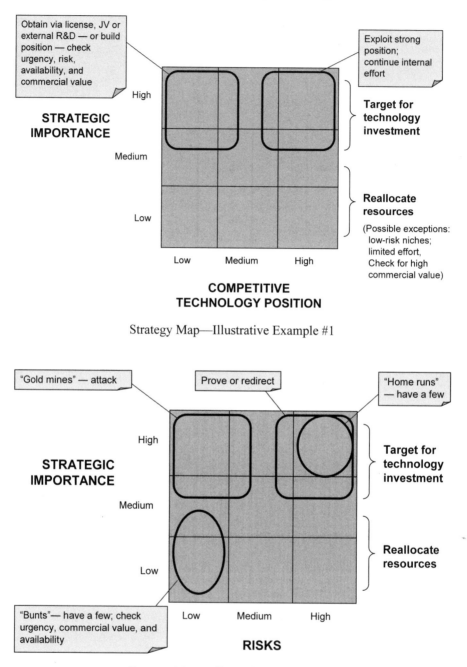

Strategy Map—Illustrative Example #1

Strategy Map—Illustrative Example #2

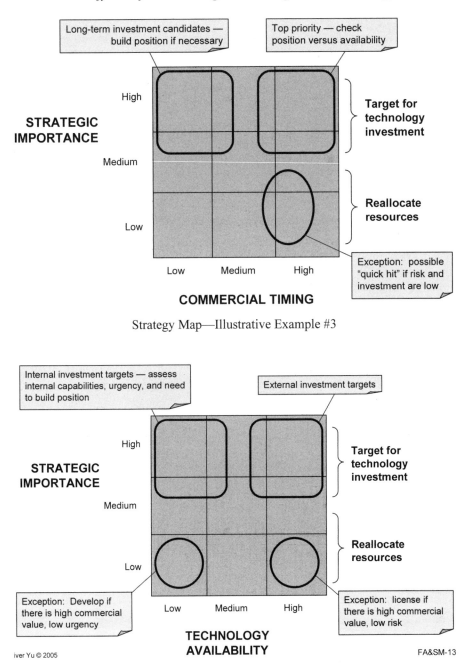

Strategy Map—Illustrative Example #3

Strategy Map—Illustrative Example #4

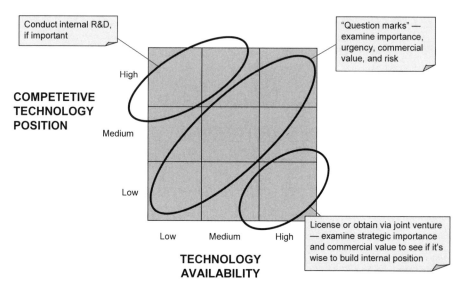

Strategy Map—Illustrative Example #5

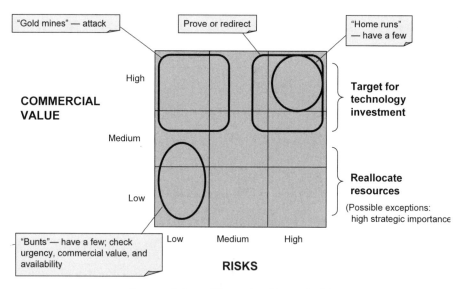

Strategy Map—Illustrative Example #6

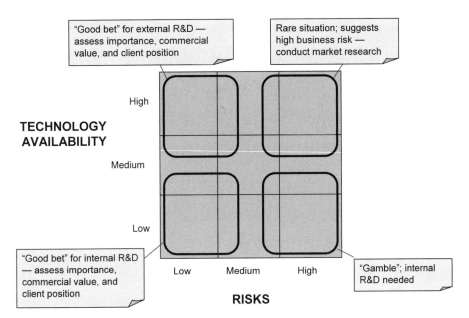

Strategy Map—Illustrative Example #7

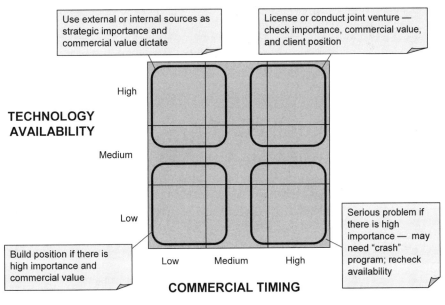

Strategy Map—Illustrative Example #8

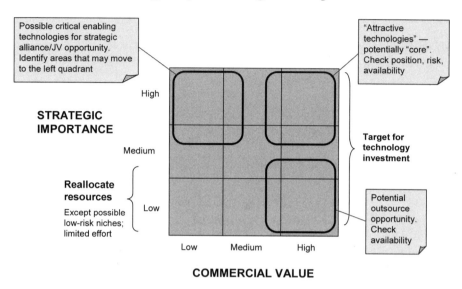

Strategy Map—Illustrative Example #9

9.2.2. Position technologies under different scenarios

With the strategy map, we can position technologies in an existing or proposed portfolio under different scenarios. The following figures provide some typical examples for three strategy maps for a portfolio of 15 technologies evaluated under three scenarios taken from an actual application to technology portfolio strategy development for a major U.S. corporation.

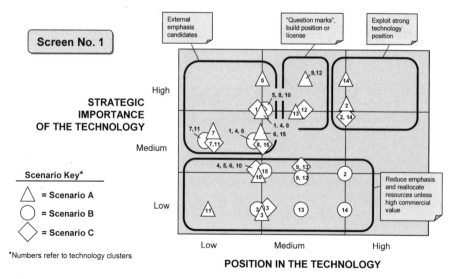

Qualitative Portfolio Evaluation—Example

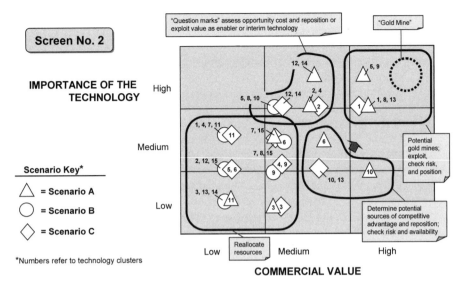

Qualitative Portfolio Evaluation—Example *(continued)*

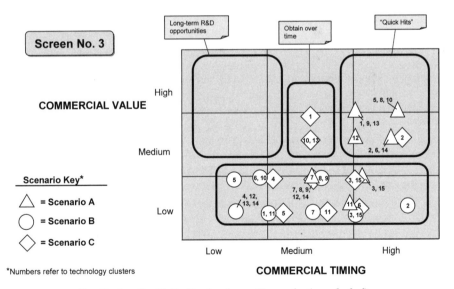

Qualitative Portfolio Evaluation—Example *(concluded)*

These positions of technologies in the strategy maps indicate the strengths, weaknesses, and responsiveness of the portfolio to various scenarios and the associated organizational strategies. The strategy maps can further provide insights about how the portfolio may be *strengthened and improved*.

As an illustrative example, the above three strategic maps provide the following preliminary insights about the technology portfolio:

- Technologies 1 and 8 appear well-positioned as candidates for "core" technologies.
- Nine technologies, 2, 3, 4, 5, 6, 12, 13, 14, 15, appear "attractive" in two or more scenarios and best represented in Scenario A.
- The organization has a competitive position in technologies 2, 12, 13, and 14; how to improve the competitive position of technologies 1, 3, 4, 5, 6, and 8 should be studied.
- Technologies with low importance and low values, especially technologies 7, 9, 10, and 11, need to be reviewed for potential downsizing or elimination.
- Studies should be made to see how the portfolio could have a few more "bunts," "home runs," and "gold mines."
- The more important technologies all have medium to high availability—a key assumption to be reviewed for accuracy.

9.3. Modify Technology Portfolio to Reintegrate with Organizational Strategy

The results of the integrated factor analysis together with the distribution of the technologies of the selected portfolio in various areas of the strategy maps provide the decision maker with two types of insights.

First, these results reveal how much the technology portfolio selected through a systematic disciplined planning process is consistent with and responsive to the organization's visions of the future and the current corporate strategy. Any mismatches would indicate that gaps exist in the responsiveness of the portfolio to the planning scenarios and the ability of the portfolio to effectively support the organizational strategy. This mismatch provides the driving values for the revisions of the planned technology portfolio selection. These revisions will not only further align the technology portfolio with the organizational strategy, but also refine the technology portfolio planning process by identifying potential misunderstanding or misinterpretation of the corporate values and technology requirements based on the planning scenarios and the organizational strategy.

Second, as the organizational strategy evolves with changes in the external business environment as well as progression in internal technology development or application, the results from the integrated factor analysis and strategy maps reveal how much the underlying assumptions on values, alternatives, and relationships of the portfolio planning process have shifted. These shifts provide the basis for the decision maker to take appropriate actions to effectively manage the current technology portfolio by strengthening, keeping constant, reducing, or eliminating the resources allocated to existing technologies in the portfolio and adding resources to new technologies to be included in the portfolio. Through these modifications, the technology portfolio will be reintegrated into the evolving organizational strategy to provide continuing support to the changing organizational goals and objectives.

Finally, it is useful to note that the strategy maps, particularly those related to Strategic Importance and Commercial Value versus Risks and Position, can be and have been used as effective qualitative tools for developing insights about technology portfolio diversification and optimization following the principles discussed in Chapter 7.

9.4. References

Henderson, B., *The Logic of Business Strategy*, Ballinger, 1984

Hermann, K., *Visualizing Your Business: Let Graphics Tell the Story*, Wiley, 2001

Daichendt, G. and Johnson, B., *I-Operations: The Impact of the Internet on Operating Models*, The Institute Press, 2000

9.5. Exercises

Problem 9.1

For the technology portfolio in your business or personal life, such as various types of computer, telecommunications, and video equipment your company has or you personally have, use the factor analysis technique to assess its strengths and weaknesses.

Problem 9.2

For the technology portfolio in Problem 9.1, develop two major future scenarios of interest to you and use the strategy map technique to assess the adequacy of the portfolio in effectively responding to these different scenarios and suggest possible modifications to the portfolio.

Appendix   Basic Mathematical, Probability, and Statistical Concepts and Tools

This appendix provides a summary of the basic mathematical, probabilistic, and statistical concepts and tools that are prerequisites for many of topics discussed in the book.

A1. Basic Mathematical Concepts and Tools

A1.1. Differentiation and local optima

For a function $f(x)$ that is continuous and twice differentiable, then the local maximum occurs at the value x^* for which the first derivative of f with respect to x, $df(x)/dx = 0$, and the second derivative, $d^2f(x)/dx^2 < 0$; on the other hand, the local minimum occurs at the value x^* for which $df(x)/dx = 0$ and $d^2f(x)/dx^2 > 0$.

A1.2. Matrix theory and operations

A *matrix* $A=[a_{ij}]$ is a rectangular array of m rows and n columns of numbers, with aij being the number at row i and column j.

$A^T=[a_{ji}]$, which is the *transpose* of A, is the nxm matrix, with its columns being the rows of $A=[a_{ij}]$.

A *vector* $x=[x_i]$ is a mx1 matrix and $x>0$ mean all $x_i>0$.

A mxp matrix $A=[a_{ij}]$ can be multiplied to a pxn matrix $B=[b_{jk}]$ to produce a mxn matrix $C=[c_{ik}]$ with $c_{ik}=\Sigma_j\, a_{ij}b_{jk}$.

The *identity* matrix I is an nxn matrix with all diagonal elements being 1 and all other elements being 0.

The *zero* matrix is an nxn matrix of 0s.

A nxn matrix A is *singular* if one of its columns, A_i, is the linear combination of other columns, A_j and A_k, i.e., $A_i=aA_j + bA_k$, where a and b are constants; otherwise A is *non-singular*.

For an nxn matrix A, if there exists an nx1 non-zero vector x so that $Ax=\lambda x$ for some constant value λ, then λ is the eigenvalue of A for the corresponding eigenvector x. λ is also the root of the polynomial equation obtained by setting the determinant of $(A-\lambda I)=0$, where I is the nxn identify matrix.

A2. Basic Probability Concepts and Tools

A2.1. Probability definitions and rules

An *event* is the outcome of an experiment, and the simplest event is called an *elementary event*; all events can be represented by collections of elementary events.

The event representing the non-occurrence of an event A is called the *complement* of A and labeled as A^c or not A; the collection of all possible events is the *sample space* S, and the complement of S is the *empty* event.

Events are *collectively exhaustive* if the collection of these events is the sample space; as a special case, an event and its complement are collectively exhaustive; two events are *mutually exclusive* if their intersection is the empty event; all elementary events as well as an event and its complement are mutually exclusive.

The *intersection* of two events A and B, AB, is the collection of elementary events that belongs to both A and B.

The *union* of two events A and B, AUB, is the collection of elementary events that belongs to either A or B or both; if A_i, i=1,2,...,n are mutually exhaustive and collectively exhaustive, then the union of all A_i, $UA_i = A_1UA_2U...UA_n = S$, and the intersection of UA_i with event B, $(UA_i)B$ is the union of A_iB for all A_i and equals B.

A *probability* is a numerical measure of the chance of occurrence of an event with the following basic characteristics:

1. Each probability lies between 0 and 1
2. The probability of the sample space is 1

As a result, $P(A)=1-P(A^c)$.

The probability of a single event is also called the *marginal* probability and the probability of the intersection of two events A and B is called the *joint* probability P(AB).

The Addition Rule: For two events A and B, P(AUB)=P(A)+P(B)-P(AB); if A and B are mutually exclusive then P(AUB) = P(A)+P(B).

Conditional probability, P(A|B), of probability of A given the condition that B has already occurred, is defined to be the ratio of the joint probability P(AB) and the marginal probability of the condition B, i.e., P(A|B) = P(AB)/P(B), provided that P(B) is positive.

Two events A and B are *independent* if P(A|B) = P(A) and P(B|A) = P(B).

Multiplication Rule: For two events A and B, P(AB)=P(A|B)/P(B)=P(BA)= P(B|A)P(A); if A and B are independent, then P(AB)=P(A)P(B); specifically if A_i, i=1,2,...,n are mutually exhaustive and collectively exhaustive, then $P(B)=P[(UA_i)B]=P[U(A_iB)]=\Sigma_iP(A_iB)=\Sigma_iP(BA_i)$

A2.2. Probability assignment

Assigning probabilities to events is complex and often difficult. There are three basic approaches to probability assignment, but unfortunately they are flawed:

1. *Classical approach* which assumes that all elementary events are equally likely; clearly this assumption is not generally true in reality
2. *Experimental or relative frequency approach* which assigns probability to an event to be the relative frequency of occurrence of the event during an experiment; this approach is experiment-dependent and often does not yield a precise assignment
3. *Subjective approach* which argues that in a decision problem only the decision maker's subjective judgment about the probability of an event matters; this approach suffers from the many inadequacies and contradictions of human judgments

Given these flaws, we need to use a combination of all three approaches to develop the best possible probability assignment of a given event.

A2.3. Bayes Theorem

Prior probabilities $P(A_i)$s of mutually exclusive and collectively exhaustive event A_i, $i=1,2,\ldots,n$ is the respective subjective probabilities of a decision maker assigns to the occurrence of these events prior to observing any new evidence.

The *posterior probability* $P(A_i|B)$ is the probability of event A_i given that a new evidence B has been observed by the decision maker.

If the conditional probabilities $P(B|A_i)$ for $i=1,2,\ldots,n$ are known from past history or experimental results, then the *Bayes Theorem* can be derived as follows:

$$P(A_i|B) = P(A_iB)/P(B) = P(BA_i)/ \Sigma_j P(BA_j) = P(B|A_i)P(A_i)/ \Sigma_j P(B|A_j)P(A_j)$$

Bayes Theorem provides a logical, systematic, and precise procedure for revising the prior probabilities by new evidence.

A3. Basic Statistical Concepts and Tools

A3.1. Population data and parameters

The *population* is an entire collection of available data, and the *parameters* are special characteristics of the population data distribution that are of particular interest to the decision maker.

There are two major types of data:

1. *qualitative* data that are descriptors, such as color, gender, etc., and are not mathematically operable; for this type of data, the parameters of interest are the respective *proportions* p_i's of the descriptor groups.
2. *quantitative* data that are mathematically operable; for a population of N data points x_i, $i=1,2,\ldots,N$, two major parameters are:

    Population mean $= \mu = \Sigma x_i/N$
    Population variance $= \sigma^2 = \Sigma (x_i-\mu)^2/N$

A parameter that is totally equivalent to the population variance is the *population standard deviation* σ.

It is useful to note that for the qualitative data, if there are only two descriptor groups in the population with p being the proportion of the group artificially labeled as 1 and 1–p being the proportion of the other group which is artificially labeled as 0, then these labels may be viewed as quantitative and μ and σ^2 are then p and p(1–p), respectively.

A3.2. Random sample and statistics

Because N is generally very large, the decision maker often does not have the resources or time to find μ and σ directly from the population, thus a *sample* of the population data will be collected to provide a basis for estimating the parameters.

To be representative of the population, a *random sample* must be collected so that each element in the population is equally likely to be selected. In this way, each element in the sample before it is observed would have a probability distribution for its values identical to the relative frequency distribution of the population.

The representative characteristics of the sample data distribution are called *statistics*. For a sample with n data, x_i, two major sample statistics that are also respectively estimators of μ and σ are:

Sample mean $= \overline{x} = \Sigma x_i / n$
Sample variance $= s^2 = \Sigma(x_i - \overline{x})^2 / (n-1)$

Similarly, sample variance s^2 and *sample standard deviation* s are totally equivalent.

It is useful to note that the mean and variance of a random sample have probability distributions with the following characteristics:

Expected value of the sample mean $= \mu$
Expected value of the sample variance $= \sigma^2 / n$

This latter characteristic is the basis of the well-known *Law of Large Numbers*, which states that as sample size n increases, sample mean approaches population mean.

A3.3. Central limit theorem, normal and student t distributions

To estimate the value of the population mean with *confidence*, we need to know the probability distribution of the sample mean, or the *sampling distribution*, which is generally difficult to determine for an arbitrary population data distribution. Fortunately, a universal law, the *Central Limit Theorem*, exists, so that regardless of the population distribution, for a random sample size n≥30, the sample mean will have an approximately normal probability distribution with mean equal μ and variance equal to σ^2 / n.

A general version of the Central Limit Theorem states that the sum of a large number of independent random variables is approximately normally distributed with its mean equal to the sum of the individual means and its variance equal to the sum of the individual variances.

On the other hand, in the special case where the population is known to be approximately normal but the population variance is unknown, then for a random sample of any size n, $\bar{x}/s$ will have a student t distribution of n–1 degree of freedom.

Standard normal distribution and student t distribution tables are given at the end of the Appendix.

A3.4. Interval estimation and hypothesis testing

In *interval estimation*, we want to find the interval $\bar{x} \pm E$ that has a prescribed 1–α level of probability or confidence of covering the population mean.

It can be readily shown that if the sampling distribution is:

1. normal, then $E = z_{\alpha/2}\,(s/\sqrt{n}\,)$.
2. t-distribution, then $E = t_{n-2,\,\alpha/2}\,(s/\sqrt{n}\,)$

where $z_{\alpha/2}$ and $t_{n-1,\,\alpha/2}$ are, respectively, the values of the standard normal random variable and the student t random variable at which the tail-end probability is equal to α/2.

Similarly, in *hypothesis testing* with a statistical significance α, the null hypothesis Ho: $\mu \geq \mu o$ will be rejected if the sampling distribution is:

1. normal, and $\bar{x} < \mu o - z_{\alpha/2}\,s/\sqrt{n}$
2. t-distribution, and $\bar{x} < \mu o - t_{n-2,\,\alpha/2}\,s/\sqrt{n}$

A3.5. Linear regression analysis

For linear regression analysis for the relationship between a dependent variable y and a single independent variable x, it is assumed that

$$y = \alpha + \beta x + \varepsilon$$

where ε is a random error in measurements or recording, generally assumed to be normally distributed with mean 0 and standard deviation σ.

For this linear equation, since α and β are generally unknown, we will need to estimate them respectively from the observed x_i and y_i through the following estimated relationship:

$$y_i = a + bx_i,$$

Specifically, linear regression analysis is used to estimate the best-fit values for a and b that minimize the sum of the squares of the errors between the observed y_i and the estimated value $a+bx_i$ (hence these coefficients are often called the least square best fit estimates).

Formulas for the Least Square Estimates of the Slope b and Intercept a are summarized below:

$$b = [\Sigma x_i y_i - (\Sigma x_i \, \Sigma y_i)/n]/[\Sigma x_i^2 - (\Sigma x_i)^2/n]$$
$$a = (\Sigma y_i/n) - b(\Sigma x_i/n)$$

where n = number of historical time periods

Under the assumption that the random errors at different times have independent, identical normal distributions with mean 0 and standard deviation σ, the ratio, b/s, between the estimated slope b and the sample standard deviation of the statistical error s, has a student t distribution of $n-2$ degree of freedom. It can then be used to test the null hypothesis, Ho: $\beta=0$. Specifically, at a statistical significance α, Ho will be rejected if $b/s > t_{n-2, \, \alpha/2}$.

Furthermore, because the random error is normally distributed, we have the $1-\alpha$ confidence interval for

1. the mean of all y's corresponding to $x=x_o$ as:

$$a+bx_o \pm t_{n-2, \, \alpha/2} (s_{y.x}) \sqrt{[(1/n) + x_o - \overline{x})^2 / \Sigma \cdot (x_i - \overline{x})^2]}$$

2. (2) the mean of the y corresponding to a specific $x=x_o$ as:

$$a+bx_o \pm t_{n-2, \, \alpha/2} (s_{y.x}) \sqrt{[1 + (1/n) + x_o - \overline{x})^2 / \Sigma \cdot (x_i - \overline{x})^2]}$$

where $s_{y.x} = \sqrt{[\Sigma \cdot (y_i - a - bx_i)^2 / (n-2)]}$

Table A-1. Standard Normal Distribution Table: Probability Content from $-\infty$ to Z

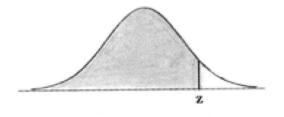

z	0.00	0.01	0.02	0.03	0.04	0.05	0.06	0.07	0.08	0.09
0.0	0.5000	0.5040	0.5080	0.5120	0.5160	0.5199	0.5239	0.5279	0.5319	0.5359
0.1	0.5398	0.5438	0.5478	0.5517	0.5557	0.5596	0.5636	0.5675	0.5714	0.5753
0.2	0.5793	0.5832	0.5871	0.5910	0.5948	0.5987	0.6026	0.6064	0.6103	0.6141
0.3	0.6179	0.6217	0.6255	0.6293	0.6331	0.6368	0.6406	0.6443	0.6480	0.6517
0.4	0.6554	0.6591	0.6628	0.6664	0.6700	0.6736	0.6772	0.6808	0.6844	0.6879
0.5	0.6915	0.6950	0.6985	0.7019	0.7054	0.7088	0.7123	0.7157	0.7190	0.7224
0.6	0.7257	0.7291	0.7324	0.7357	0.7389	0.7422	0.7454	0.7486	0.7517	0.7549
0.7	0.7580	0.7611	0.7642	0.7673	0.7704	0.7734	0.7764	0.7794	0.7823	0.7852
0.8	0.7881	0.7910	0.7939	0.7967	0.7995	0.8023	0.8051	0.8078	0.8106	0.8133
0.9	0.8159	0.8186	0.8212	0.8238	0.8264	0.8289	0.8315	0.8340	0.8365	0.8389
1.0	0.8413	0.8438	0.8461	0.8485	0.8508	0.8531	0.8554	0.8577	0.8599	0.8621
1.1	0.8643	0.8665	0.8686	0.8708	0.8729	0.8749	0.8770	0.8790	0.8810	0.8830
1.2	0.8849	0.8869	0.8888	0.8907	0.8925	0.8944	0.8962	0.8980	0.8997	0.9015
1.3	0.9032	0.9049	0.9066	0.9082	0.9099	0.9115	0.9131	0.9147	0.9162	0.9177
1.4	0.9192	0.9207	0.9222	0.9236	0.9251	0.9265	0.9279	0.9292	0.9306	0.9319
1.5	0.9332	0.9345	0.9357	0.9370	0.9382	0.9394	0.9406	0.9418	0.9429	0.9441
1.6	0.9452	0.9463	0.9474	0.9484	0.9495	0.9505	0.9515	0.9525	0.9535	0.9545
1.7	0.9554	0.9564	0.9573	0.9582	0.9591	0.9599	0.9608	0.9616	0.9625	0.9633
1.8	0.9641	0.9649	0.9656	0.9664	0.9671	0.9678	0.9686	0.9693	0.9699	0.9706
1.9	0.9713	0.9719	0.9726	0.9732	0.9738	0.9744	0.9750	0.9756	0.9761	0.9767
2.0	0.9772	0.9778	0.9783	0.9788	0.9793	0.9798	0.9803	0.9808	0.9812	0.9817
2.1	0.9821	0.9826	0.9830	0.9834	0.9838	0.9842	0.9846	0.9850	0.9854	0.9857
2.2	0.9861	0.9864	0.9868	0.9871	0.9875	0.9878	0.9881	0.9884	0.9887	0.9890
2.3	0.9893	0.9896	0.9898	0.9901	0.9904	0.9906	0.9909	0.9911	0.9913	0.9916
2.4	0.9918	0.9920	0.9922	0.9925	0.9927	0.9929	0.9931	0.9932	0.9934	0.9936
2.5	0.9938	0.9940	0.9941	0.9943	0.9945	0.9946	0.9948	0.9949	0.9951	0.9952
2.6	0.9953	0.9955	0.9956	0.9957	0.9959	0.9960	0.9961	0.9962	0.9963	0.9964
2.7	0.9965	0.9966	0.9967	0.9968	0.9969	0.9970	0.9971	0.9972	0.9973	0.9974
2.8	0.9974	0.9975	0.9976	0.9977	0.9977	0.9978	0.9979	0.9979	0.9980	0.9981
2.9	0.9981	0.9982	0.9982	0.9983	0.9984	0.9984	0.9985	0.9985	0.9986	0.9986
3.0	0.9987	0.9987	0.9987	0.9988	0.9988	0.9989	0.9989	0.9989	0.9990	0.9990

Table A-2. Student t Distribution Table: Value for Tail-end Probability α

df	$\alpha = 0.1$	0.05	0.025	0.01	0.005	0.001	0.0005
1	3.078	6.314	12.706	31.821	63.656	318.289	636.578
2	1.886	2.920	4.303	6.965	9.925	22.328	31.600
3	1.638	2.353	3.182	4.541	5.841	10.214	12.924
4	1.533	2.132	2.776	3.747	4.604	7.173	8.610
5	1.476	2.015	2.571	3.365	4.032	5.894	6.869
6	1.440	1.943	2.447	3.143	3.707	5.208	5.959
7	1.415	1.895	2.365	2.998	3.499	4.785	5.408
8	1.397	1.860	2.306	2.896	3.355	4.501	5.041
9	1.383	1.833	2.262	2.821	3.250	4.297	4.781
10	1.372	1.812	2.228	2.764	3.169	4.144	4.587
11	1.363	1.796	2.201	2.718	3.106	4.025	4.437
12	1.356	1.782	2.179	2.681	3.055	3.930	4.318
13	1.350	1.771	2.160	2.650	3.012	3.852	4.221
14	1.345	1.761	2.145	2.624	2.977	3.787	4.140
15	1.341	1.753	2.131	2.602	2.947	3.733	4.073
16	1.337	1.746	2.120	2.583	2.921	3.686	4.015
17	1.333	1.740	2.110	2.567	2.898	3.646	3.965
18	1.330	1.734	2.101	2.552	2.878	3.610	3.922
19	1.328	1.729	2.093	2.539	2.861	3.579	3.883
20	1.325	1.725	2.086	2.528	2.845	3.552	3.850
21	1.323	1.721	2.080	2.518	2.831	3.527	3.819
22	1.321	1.717	2.074	2.508	2.819	3.505	3.792
23	1.319	1.714	2.069	2.500	2.807	3.485	3.768
24	1.318	1.711	2.064	2.492	2.797	3.467	3.745
25	1.316	1.708	2.060	2.485	2.787	3.450	3.725
26	1.315	1.706	2.056	2.479	2.779	3.435	3.707
27	1.314	1.703	2.052	2.473	2.771	3.421	3.689
28	1.313	1.701	2.048	2.467	2.763	3.408	3.674
29	1.311	1.699	2.045	2.462	2.756	3.396	3.660
30	1.310	1.697	2.042	2.457	2.750	3.385	3.646
60	1.296	1.671	2.000	2.390	2.660	3.232	3.460
120	1.289	1.658	1.980	2.358	2.617	3.160	3.373
∞	1.282	1.645	1.960	2.326	2.576	3.091	3.291

Index

n denotes an entry located within a footnote.